Lasers and Electro-Optics Research and Technology

QUANTUM WELL STRUCTURES FOR INFRARED PHOTODETECTION

LASERS AND ELECTRO-OPTICS RESEARCH AND TECHNOLOGY

Phase Transitions Induced by Short Laser Pulses
Georgy A. Shafeev (Editor)
2009. 978-1-60741-590-9

Performance Monitoring in Advanced Optical Fiber Networks
Lianshan Yan (Author)
2009. 978-1-60692-645-1

**High-Power and Femtosecond Lasers:
Properties, Materials and Applications**
Paul-Henri Barret and Michael Palmer (Editors)
2009. 978-1-60741-009-6

From Femto-to Attoscience and Beyond
*Janina Marciak-Kozlowska
and Miroslaw Kozlowski (Authors)*
2009. 978-1-60741-164-2

Fiber Lasers: Research, Technology and Applications
Masato Kimura (Editor)
2009. 978-1-60692-896-7

**Optical Solitons in Nonlinear Micro Ring Resonators:
Unexpected Results and Applications**
*Nithiroth Pornsuwancharoen, Jalil Ali
and Preecha Yupapin (Authors)*
2009. 978-1-60741-342-4

Fiber Lasers: Research, Technology and Applications
Masato Kimura (Editor)
2009. 978-1-60876-777-9

**High-Power and Femtosecond Lasers:
Properties, Materials and Applications**
Paul-Henri Barret and Michael Palmer (Editors)
2009. 978-1-60876-739-7

Quantum Well Structures for Infrared Photodetection
Wei Shi and D.H. Zhang (Authors)
2010. 978-1-61668-368-9

Laser Beams: Theory, Properties and Applications
Maxim Thys and Eugene Desmet (Editors)
2010. 978-1-60876-266-8

**Optical and Electro-Optical Properties of Liquid Crystals:
Nematic and Smecic Phases**
Minko Parvanov Petrov (Author)
2010. 978-1-61668-360-3

Quantum Well Structures for Infrared Photodetection
Wei Shi and D.H. Zhang (Authors)
2010. 978-1-61668-827-1

Lasers and Electro-Optics Research and Technology

QUANTUM WELL STRUCTURES FOR INFRARED PHOTODETECTION

WEI SHI
AND
D.H. ZHANG

Nova Science Publishers, Inc.
New York

Copyright © 2010 by Nova Science Publishers, Inc.

All rights reserved. No part of this book may be reproduced, stored in a retrieval system or transmitted in any form or by any means: electronic, electrostatic, magnetic, tape, mechanical photocopying, recording or otherwise without the written permission of the Publisher.

For permission to use material from this book please contact us:
Telephone 631-231-7269; Fax 631-231-8175
Web Site: http://www.novapublishers.com

NOTICE TO THE READER

The Publisher has taken reasonable care in the preparation of this book, but makes no expressed or implied warranty of any kind and assumes no responsibility for any errors or omissions. No liability is assumed for incidental or consequential damages in connection with or arising out of information contained in this book. The Publisher shall not be liable for any special, consequential, or exemplary damages resulting, in whole or in part, from the readers' use of, or reliance upon, this material.

Independent verification should be sought for any data, advice or recommendations contained in this book. In addition, no responsibility is assumed by the publisher for any injury and/or damage to persons or property arising from any methods, products, instructions, ideas or otherwise contained in this publication.

This publication is designed to provide accurate and authoritative information with regard to the subject matter covered herein. It is sold with the clear understanding that the Publisher is not engaged in rendering legal or any other professional services. If legal or any other expert assistance is required, the services of a competent person should be sought. FROM A DECLARATION OF PARTICIPANTS JOINTLY ADOPTED BY A COMMITTEE OF THE AMERICAN BAR ASSOCIATION AND A COMMITTEE OF PUBLISHERS.

LIBRARY OF CONGRESS CATALOGING-IN-PUBLICATION DATA

Available upon Request
ISBN: 978-1-61668-368-9

Published by Nova Science Publishers, Inc. ✢ New York

Contents

Preface		ix
Chapter 1	Introduction	1
Chapter 2	Theoretical Review	5
Chapter 3	P-Doped GaInAs/AlGaAs Strained MQW Structures	15
Chapter 4	Quantum Well Infrared Photodetectors	49
Chapter 5	Conclusion	63
Acknowlegements		65
References		67
Index		73

PREFACE

Electrical, optical and structural properties of the Be-doped GaInAs/AlGaAs multiple quantum-well structures (MQWs) at different doping density in the GaInAs well, designed and fabricated in house, were extensively studied. The higher Be-doping density in the wells was found to enhance the compressive strain and increase barrier height of the GaInAs/AlGaAs MQWs through extensive optical characterization. It caused the shifts of the sub-energy levels in the valence band of the well material and the absorption wavelength resulting from the intersubband absorption. These observations were verified by our theoretical calculation based on the six-band Luttinger-Kohn model by taking the Be-doping into account.

The above compressively strained InGaAs/AlGaAs MQWs have been fabricated into quantum-well infrared photodetectors (QWIP) devices. The dark current of the devices was found to be much lower than that unstrained GaAs/AlGaAs MQWs, and two different mechanisms of conduction were distinguished to dominate at the two different temperature ranges. The device also shows a strong phototoresponse peaked at 6.2 μm with a Be concentration of 10^{18} cm^{-3} in the wells. Its responsivity and blackbody detectivity are symmetric for forward and reversed bias, and comparable to and even better than some p-type QWIPs made of other material systems. By increasing Be doping in the wells, the detected peak wavelength becomes smaller and the responsivity and detectivity become asymmetric due to the bandgap narrowing at high doping and inhomogeneity near the well-barrier interfaces.

Chapter 1

1. INTRODUCTION

At room temperature, an object emits most of its energy in the form of infrared (IR) radiation. Therefore, in observing terrestrial objects and activities without light reflection, infrared radiation is one of the most important spectra to be monitored. Infrared detectors classified as either thermal or photon devices can be thought as transducers that convert IR radiation into an electrical signal [1]. In photon detectors, the absorption of long-wavelength radiation results directly in some specific quantum events, such as photoelectric emission of the electrons from a surface, or electronic interband transitions in semiconductor materials. In the case of photon detection, it is not necessary to heat the entire device material to sense the radiation, as is necessary in thermal detectors. Hence, these detectors are much faster than thermal detectors.

However, finding a suitable material for infrared detection is always a challenge. With the maturing of the compound semiconductor technologies (in particular, the Gallium Arsenide technology), compound semiconductor materials with various bandgaps become good candidates for the desirable optical absorption. In recent years, major steps have been made toward a new approach for designing semiconductor structures with tailored electronic and optical properties for a new generation of long-wavelength infrared quantum detectors. This approach has come to be known as bandgap engineering [2,3].

Conventional interband optical absorption involves photoexciting carriers across the band gap E_g, i.e., promoting an electron from the valence-band ground state to the conduction-band excited state. In a detector structure, these photocarriers are collected, thereby producing a photocurrent. By controlling E_g, the spectrum of the absorption and hence the wavelength dependence of the detector response can be modified. That means long wavelength detection needs a photodetector based on small-band-gap materials. However these kinds of materials are more difficult to grow, process, and fabricate into devices than the larger-band-gap ones. These difficulties thus motivate the studies of novel "artificial" low "effective" band-gap ($E_g < 1$ eV) semiconductors. In particular, semiconductor multiple quantum well (MQW) structures have been attracting considerable attention in the recent years. They can reach infrared spectral region using intersubband (i.e., intraband) absorption. These detectors so-called Quantum Well Infrared Photodetectors (QWIPs) are based on the optical transition within a single energy band (intersubband transition) and are therefore independent on the bandgap of the detecting material. Based on the new detector concept, a material can be used to detect a much lower energy radiation than what is allowed by the separation between its valence band and the conduction band. The choice of detector material is thus substantially widened, and in most cases, the conventional material can be substituted by a more mature and better understood material system.

In recent years, the III-V compound semiconductor MQW structures have achieved high-detectivity performance so that the large two-dimensional staring arrays (128×128) with long-wavelength infrared imaging performance become comparable to state-of-the-art HgCdTe. The QWIP arrays have a number of advantages including the use of standard manufacturing techniques based on the mature GaAs growth and processing technologies (currently used for large-scale high-speed integrated circuits, lasers, microwave circuits, etc.), highly uniform and well-controlled molecular beam epitaxy (MBE) or metalorganic chemical vapor deposition (MOCVD) growth (leading to excellent array uniformity), high-yield (and thus low-cost), monolithic integration with GaAs signal processing electronics, intrinsic radiation hardness, and multispectral detection. The material uniformities are typically an order of magnitude better than those for HgCdTe and higher uniformity means higher performance [4]. A further advantage, as far as fabrication and producibility are concerned, is that very simple mesa processing can be used. In particular, no passivation is required. Both III-V QWIPs and HgCdTe technologies can achieve a high noise-equivalent flux density. For earth-sensing satellite systems, extended long-wavelength response ($\lambda > 14\mu m$) is essential to measure CO_2 and other molecular absorption lines [5]. QWIPs have already demonstrated excellent performance in this wavelength region [6]. Furthermore, as the operating wavelength gets longer it is more and more difficult to produce high-uniformity HgCdTe, while GaAs based QWIPs remain highly uniform (typically a few percent in λ_c across the array). Thus, QWIPs are quite promising for applications requiring $\lambda>14\mu m$. Finally, for very low-background space borne surveillance systems (where space rather than the earth generates the background thermal flux), high detectivity is the most important parameter and thus HgCdTe is superior.

The first proposal and experimental result for infrared detection in III-V quantum well structures were reported by Chiu et al. [7] and Smith et al. [8] in 1983. They observed photocurrent in an n-type GaAs/AlGaAs MQW device that was illuminated with a broadband IR source. The mechanism proposed for infrared absorption was phonon-assisted free-carrier absorption in the GaAs wells. They also proposed a valence-intersubband absorption QWIP detector consisting of InAlAs wells and InP barriers. Coon and Karunasiri [9] proposed in 1984 an IR detector based on photoemission from an asymmetric undoped single QW. In this design, electrons are injected into the well by an applied electric field, and the "trapped" carriers are photoemitted from the well. Goossen and Lyon [10] followed this by proposing a grating-enhanced detector to obtain high collection efficiency with normally incident radiation. In 1985, West and Eglash observed strong intersubband (envelope wavefuction) transition in $GaAs/Al_xGa_{1-x}As$ multiple quantum wells (MQWs) that initiated serious investigations in using MQWs as infrared photodetectors. In 1987, Choi et al. and Levine et al. published a series of papers illustrating the basic operating principles of the detector, and demonstrating first QWIP based on this transition [11]. Further performance improvements were achieved by Levine et al. [12] by using transitions from the ground state to "resonant" states in the continuum. In 1991, Bethea et al. obtained the first infrared image using a 10-element linear scanning array.

All of the achievements above-mentioned are for n-type quantum-well structures, which only allow intersubband transition when the optical field is parallel to the quantum-well growth direction (designated the z polarization) [13,14]. This is due to the fact that electron motion in the quantum-well direction (the z direction) is necessary for the intersubband transition of a spherically symmetric valley. Such motion can only be induced by the optical field along the quantum-well direction. Unlike n-QWIPs, which are forbidden to absorb normal incidence infrared radiation without coupling assistance, p-type QWIPs offer normal incident intersubband absorption due to the mixing between the off-zone center heavy hole and light hole states [15]. The first p-type QWIP has been successfully demonstrated by Levine et al [16] and also p-doped GaInAs/InP gives the wavelength response of 2.7 μm by S.D. Gunapala [17], other p-type QWIPs using valence intersubband transitions in the lattice-matched GaAs/AlGaAs and GaInAs/InAlAs material systems [18-22] have also been demonstrated recently. There have been several theoretical [23-25] and experimental [26,27] research efforts on the physics and application of normal-incidence long-wavelength intersubband optical transitions in different p-type material systems to QWIPs. Recently, the strained material systems have proven to be highly interesting for both fundamental physics [28,29] and for applications [30] because of the possibility of designing the valence band structure as a function of strain [31]. The use of intentionally strained heterostructures has greatly enhanced the performance of electrical and electro-optical semiconductor devices [32]. Among them, the p-type strained GaInAs/AlGaAs multiple quantum well material combination becomes particularly important for device applications in the near and far infrared wavelength range. Chu and Li [33] have studied strained p-type QWIPs of GaInAs/GaAs, GaInAs/AlGaAs, and GaInAs/AlGaAs-GaAs, and reported their superior detectivity over the unstrained p-type QWIPs. Due to the large effective mass of holes, it permits the use of much higher doping levels in the p-type multiple QWIPs. Hence there is the possibility of larger optical absorption coefficients for the p-type QWIPs than that for n-type ones, although the optical absorption coefficients is inversely proportional to the effective mass of the carriers. Meanwhile due to the large effective mass of holes, higher p-type doping level will not increase much of the thermionic and tunneling components of the dark current. However, to our best knowledge so far, there is little report on the Be-doping effects on the quantum-well structures. In this article, several groups of p-type GaInAs/AlGaAs MQWs with different Be-doping densities in the GaInAs wells grown by Solid Source Molecular Beam Epitaxy (SSMBE) system were designed, fabricated and characterized extensively. The effects of Be-doping on the electrical and optical properties, such as subband energy level, absorption wavelength, dark current and photocurrent will be discussed.

Chapter 2

2. THEORETICAL REVIEW

2.1. QUANTUM WELL PHYSICS

A typical QWIP is made of alternate layers of two different semiconductor materials. Since there is a difference in the conduction band alignment in these material layers, a series of potential barriers is formed analogous to the quantum barriers. Indeed, to the first approximation, one can ignore the underlying atomic potentials that produce the respective material band structures, and concentrate on the effects of the potential created by the band misalignment. This simplification can be brought about using envelope function approximation, in which the electrons can be treated as plane waves in free space. The effects of the atomic potentials are accommodated by substituting the mass of a free electron by the effective mass m^* of the material. It turns out that such a simplified approach is quite adequate in device design and modeling.

To calculate the band structures and the corresponding wavefunctions within envelope function approximation, besides the standard analytical method, there are also several other methods such as Wentzel-Kramers-Brillouin (WKB) approximation, the Kronig-Penney (KP) model and the transfer-matrix method. Although the KP model applies only to periodic quantum well structure, it was used here for simplification.

Schrödinger equation (2.1) is the most popular one and it usually gives satisfactory description to the energy levels in quantum wells.

$$\left[-\frac{d}{dz}\frac{1}{m^*}\frac{d}{dz}+V(z)\right]\Psi(z)=E\Psi(z), \qquad (2.1)$$

where $m^* = m_w$ in the well and $m^* = m_b$ in the barrier region.

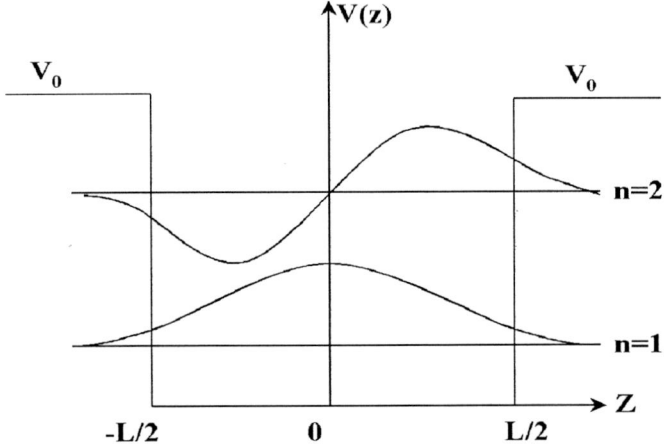

Figure 2.1. A quantum well with a width L and a finite barrier height V_0.

The energy levels for n=1 and n=2 and their corresponding wave functions For a finite barrier quantum well, as shown in figure 2.1, we have

$$V(z) = \begin{cases} V_0 & |z| \geq \dfrac{L}{2} \\ 0 & |z| < \dfrac{L}{2} \end{cases}. \qquad (2.2)$$

Using boundary conditions in which the wave function Ψ and its first derivative divided by the effective mass $(1/m^*)(d\Psi/dz)$ are continuous at the interface between the barrier and the well, that is,

$$\Psi\left(\frac{L^+}{2}\right) = \Psi\left(\frac{L^-}{2}\right)$$

and

$$\frac{1}{m_b}\frac{d}{dz}\Psi\left(\frac{L^+}{2}\right) = \frac{1}{m_w}\frac{d}{dz}\Psi\left(\frac{L^-}{2}\right). \qquad (2.3)$$

Solving the Schrödinger equation, the eigenequations or the quantization conditions for even (2.4) and odd (2.5) wave functions, respectively, were obtained.

$$\alpha = \frac{m_b k}{m_w} \tan k \frac{L}{2} \qquad (2.4)$$

$$\alpha = -\frac{m_b k}{m_w} \cot k \frac{L}{2}, \qquad (2.5)$$

where

$$k = \frac{\sqrt{2m_w E}}{\hbar} \qquad (2.6)$$

$$\alpha = \frac{\sqrt{2m_b (V_0 - E)}}{\hbar}. \qquad (2.7)$$

The eigenenergy E can be found for even and odd wave function case by searching for the root in Equation (2.4) and (2.5) respectively, together with Equation (2.6) and (2.7). The number of bound states N is determined by

$$(N-1)\frac{\pi}{2} \leq \sqrt{2m_w V_o}\left(\frac{L}{2\hbar}\right) < N\frac{\pi}{2}. \qquad (2.8)$$

The band structures of typical semiconductor quantum wells are shown as Figure 2.2.

For a given quantum well potential and in the case of conduction band, we can replace $V(z)$ with

$$V(z) = \begin{cases} V_0 (= \Delta E_C) & |z| \geq \frac{L_w}{2} \\ 0 & |z| < \frac{L_w}{2} \end{cases}, \qquad (2.9)$$

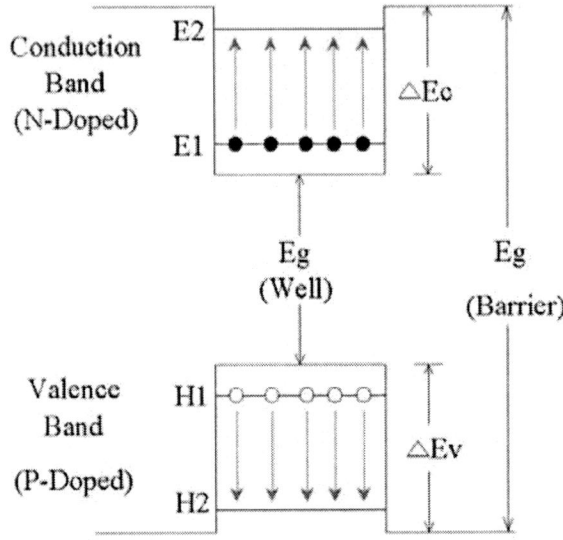

Figure 2.2. Quantum-well profiles for a semiconductor QW system.

In the case of valence band, $V(z)$ as

$$V(z) = \begin{cases} V_0(=-\Delta E_V) & |z| \geq \dfrac{L_w}{2} \\ 0 & |z| < \dfrac{L_w}{2} \end{cases}, \quad (2.10)$$

where the energies are all measured from the edge of the conduction band or valence band.

Photoelectrons excited by incident light are collected though applying bias voltage at the both terminals of the MQW structure. The valence potential profile of the MQW structure biased is illustrated in figure 2.3. The electric field can be regarded as a perturbation, hence we can also get the solutions from the Schrödinger equation with perturbation theory.

The solution to Schrödinger equation above mentioned is for a single band and the coupling to other bands is in neglected. Even with this simplicity, the results obtained here can be used to approximately estimate the band structure near the band edges.

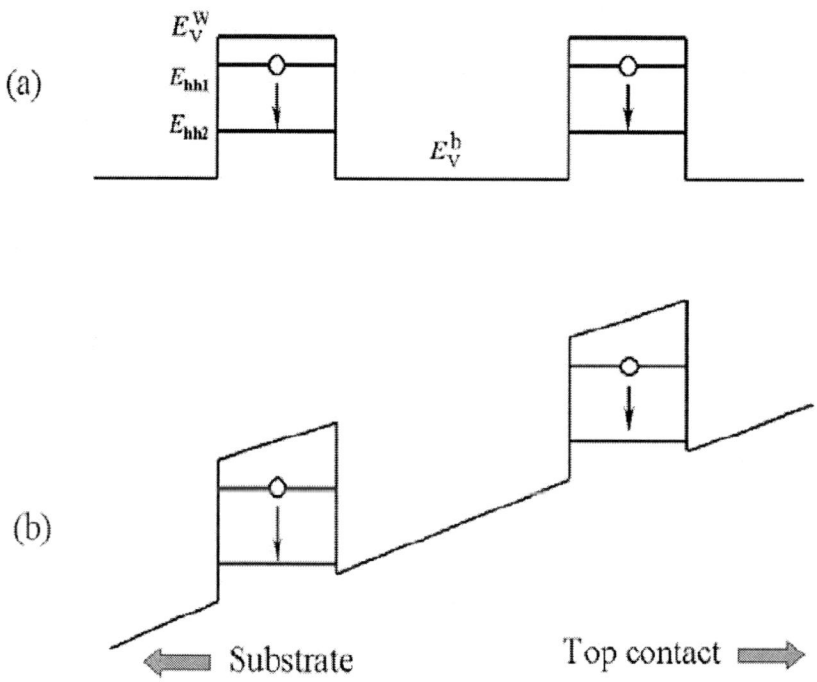

Figure 2.3. The band diagram of a double quantum well structure at (a) zero bias and (b) positive bias.

However, the interaction between the bands can also be considerd. Since most semiconductors are direct band gap materials, and many physical phenomena near the band edges (the wave vector k deviates by a small amount from a vector k_0 where a local minimum or maximum occurs) are of interest. The $\mathbf{k} \cdot \mathbf{p}$ method introduced by Bardeen [34] and Seitz [35] which is applied in conduction and valence band structures near the band edges, is extremely useful. This method has been applied to many semiconductors [36-41]. In this work, the spin-orbit interaction using Kane's model [38,39], and band degeneration using Luttinger-Kohn's models [40] were taken into consideration. These models are very useful in studying bulk and quantum-well semiconductors and have been used during the past three decades. In the p-type MQW structures, the valence bands are mainly interested. The six-valence band (the heavy-hole, light-hole, and spin-orbit split-off bands, all doubly degenerate) Luttinger-Kohn Hamiltonian H with strain effect (Equation 2.11) by Pikus-Bir was used, and it is given by [42]

$$H|j,m\rangle = -\begin{bmatrix} P+Q & -S & -\frac{1}{\sqrt{2}}S & 0 & R & \sqrt{2}R \\ -S^* & P-Q & -\sqrt{2}Q & R & 0 & \sqrt{\frac{3}{2}}S \\ -\frac{1}{\sqrt{2}}S^* & -\sqrt{2}Q & P+\Delta & -\sqrt{2}R & \sqrt{\frac{3}{2}}S & 0 \\ 0 & R^* & -\sqrt{2}R^* & P+Q & S^* & -\frac{1}{\sqrt{2}}S^* \\ R^* & 0 & \sqrt{\frac{3}{2}}S^* & S & P-Q & \sqrt{2}Q \\ \sqrt{2}R^* & \sqrt{\frac{3}{2}}S^* & 0 & -\frac{1}{\sqrt{2}}S & \sqrt{2}Q & P+\Delta \end{bmatrix} \begin{matrix} \left|\frac{3}{2},\frac{3}{2}\right\rangle \\ \left|\frac{3}{2},\frac{1}{2}\right\rangle \\ \left|\frac{1}{2},\frac{1}{2}\right\rangle \\ \left|\frac{3}{2},-\frac{3}{2}\right\rangle \\ \left|\frac{3}{2},-\frac{1}{2}\right\rangle \\ \left|\frac{1}{2},-\frac{1}{2}\right\rangle \end{matrix} \quad (2.11)$$

where

$$P = P_k + P_\varepsilon, Q = Q_k + Q_\varepsilon, R = R_k + R_\varepsilon, S = S_k + S_\varepsilon, \quad (2.12)$$

$$P_k = \left(\frac{\hbar^2}{2m_0}\right)\gamma_1(k_x^2 + k_y^2 + k_z^2),$$

$$Q_k = \left(\frac{\hbar^2}{2m_0}\right)\gamma_2(k_x^2 + k_y^2 - 2k_z^2),$$

$$R_k = \left(\frac{\hbar^2}{2m_0}\right)\sqrt{3}\left[-\gamma_2(k_x^2 - k_y^2) + 2i\gamma_3 k_x k_y\right],$$

$$S_k = \left(\frac{\hbar^2}{2m_0}\right)2\sqrt{3}\gamma_3(k_x - ik_y)k_z, \quad (2.13)$$

$$P_\varepsilon = -a_v(\varepsilon_{xx} + \varepsilon_{yy} + \varepsilon_{zz}), Q_\varepsilon = -\frac{b}{2}(\varepsilon_{xx} + \varepsilon_{yy} - 2\varepsilon_{zz}),$$

$$R_\varepsilon = \frac{\sqrt{3}}{2}b(\varepsilon_{xx} - \varepsilon_{yy}) - id\varepsilon_{xy}, \quad S_\varepsilon = -d(\varepsilon_{xz} - i\varepsilon_{yz}), \quad (2.14)$$

where $\hbar$ is the Planck constant, m_0 is the electron mass, k is the wave vector interpreted as a differential operator $-i\nabla$, ε_{ij} is the symmetric strain tensor; γ_1,

γ_2, and γ_3 are the Luttinger inverse mass parameters, a_v b, and d are the Pikus-Bir deformation potentials, Δ is the spin-orbit split-off energy, and the basis function $|j,m\rangle$ denotes the Bloch wave function. Here, the zero energy is taken to be at the top of the unstrained valence band. For the subband solution, transfer matrix method (TMM) can be used.

2.2. INTERSUBBAND TRANSITION IN QUANTUM WELLS

2.2.1. Integrated Absorption Strength for N-Type Quantum Wells

Intersubband absorption results from transitions between energy levels within the conduction or valence band (figure 2.2). To estimate the magnitude of light absorption in a MQW structure, it is convenient to use the quantum well state approach, in which all of the optical transitions are considered to be between two specific localized quantum well states. Further, the infinitely high barriers and parabolic bands are assumed here for simpicity. The energy levels in the well are given by [43]

$$E_j = \left(\frac{\hbar \pi^2}{2m^* L_w^2}\right) j^2, \qquad (2.15)$$

where L_w is the width of the quantum well, m^* is the effective mass in the well, and j is an integer. Here we consider the transition between the lowest and the first excited state. The corresponding integrated absorption strength is

$$\int_0^\infty \alpha(\nu) d(\nu) = \left(\frac{\rho_c N_w e^2 hf}{4\varepsilon_0 m^* c n_r}\right)\left(\frac{\sin^2 \theta'}{\cos \theta'}\right), \qquad (2.16)$$

where $\rho_c = N_D L_w$ is the two-dimensional density of carriers in the well, N_D is the three-dimensional carrier density, N_w is the number of doped wells, n_r is the index of refraction, θ' is the angle between the direction of the optical beam and the surface normal (inside the medium), and f is the oscillator strength given by [43]

$$f \equiv \frac{2m^*}{\hbar^2}(E_2 - E_1)<z>^2, \qquad (2.17)$$

where z is the direction normal to the quantum well. The detailed theoretical calculations on intraband transitions in n-type quantum wells can be found from the previous reports [44,45].

Since the oscillator strength only has a component along z, the optical electric field must also have a component parallel to z in order to induce an intersubband absorption. Thus, there is no appreciable intersubband absorption when the light beam propagates normal to the incidence radiation ($\theta' = 0$) to the layers. Unfortunately, most detector applications rely on this incident geometry. Therefore, much efforts have been devoted to design an efficient light coupling scheme under normal incidence. It mainly branches into four ways, namely, edge coupling, reflective gratings [46], random reflectors [47] and TIR (Total Internal Reflection) / SSD (Single-slit Diffraction) [48]. Edge coupling is accomplished by polishing a 45° facet on the substrate to allow infrared light back-illuminated onto the detector. The incident geometry is illustrated in figure 2.4. Although such a coupling scheme is useful only for single detector element or in a linear detector array, the simplicity of the fabrication procedures and the readily known intensity inside the sample makes this incident geometry a widely adopted standard for detector evaluation.

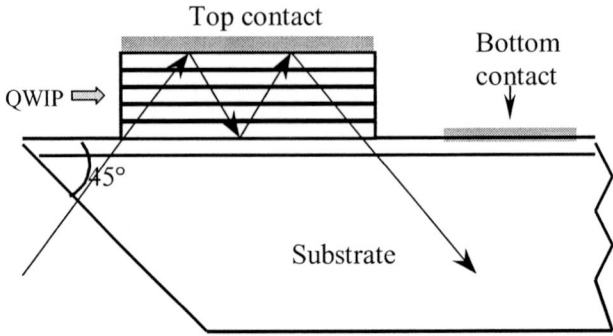

Figure 2.4. The light incident geometry under 45° edge coupling.

Besides the 45° facet incident geometry, diffraction gratings are the more efficient way of optical coupling although they introduce complicated process steps. Hasnain et al. [49] reported the first utilization of a chemically etched grating to couple light into a λ_p = 5.6 μm QWIP and the grating design was later greatly improved by Andersson et al. [50-52].

2.2.2. Intersubband Transition Occurred in the P-Type Quantum Wells

Intraband photoabsorption mentioned above is for n-type Quantum Well Structures. P-type quantum wells are more complicated because of the mixing of heavy-hole and light-hole states. The band-mixing effect has played an essential role in interpreting the $\Delta n \neq 0$ (n is the principal quantum number of quantum-well states) forbidden transitions commonly observed in the intraband absorption spectra of quantum wells [53-57]. Photoabsorption in p-type semiconductor quantum wells at photon wavelengths near Infrared is dominated by direct optical transitions between the partially filled subbands near the valence-band maximum and the unfilled higher subbands. The mechanism is similar to the intervalence-band transitions in several bulk p-type semiconductors in which direct free-carrier transitions between the heavy-hole and light-hole bands are allowed [58]. Due to the symmetry properties of valence subbands in a p-QWIP, normal-incident radiation can be detected [59] by such a device, without the assistance of waveguiding structures. Hence, p-type quantum wells are more accessible experimentally than those n-doped, because normally the intraband absorption is appreciable for incident light propagating along the growth direction.

To calculate the absorption due to intersubband transitions in p-type QWs, a suitable description of the valence band states is needed. A valence subband state $|n, k_{//}\rangle$ with in-plane wave vector $k_{//}$ can be expanded in terms of linear combinations of a set of basis states

$$\left\{ |v, k\rangle \quad v = -\frac{3}{2}, -\frac{1}{2}, \frac{1}{2}, \frac{3}{2}, \right\}, \text{ which are the Block functions,}$$

$$|n, k_{//}\rangle = \sum_{v, k_z} F_v(n k_{//}, k_z) |v, k\rangle, \qquad (2.18)$$

where n is a label for the valence subbands, and $\vec{k} = k_{//} + k_z \vec{z}$. The growth direction is along the z direction. $F_v(nk_{//}, k_z)$ can be shown to satisfy the multiband effective-mass equation,

$$\sum_{v'} \left[-H_{v,v'}^{(0)}(k_{//}, k_z) - E_n(k_{//})\delta_{v,v'} \right] F_{v'}(nk_{//}, k_z) + \sum_{k_z'} \langle k_z | V(z) | k_z' \rangle F_v(nk_{//}, k_z') = 0 \quad (2.19)$$

where $V(z)$ is the quantum well potential for the holes. The absorption coefficient for inter-valence-subband transition between subband n and n' is given by [60]

$$\alpha_{nn'}(\omega) = \frac{4\pi^2 e^2}{(\varepsilon_1)^{1/2} m_0^2 \omega c} \sum_{k_{//}} [f_n(k_{//}) - f_{n'}(k_{//})] |\varepsilon \cdot P_{nn'}(k_{//})|^2 \frac{\hbar \Gamma_{nn'}(k_{//})/\pi}{[\Delta_{nn'}(k_{//}) - \hbar\omega]^2 + [\hbar\Gamma_{nn'}(k_{//})]^2} \quad (2.20)$$

where ε_1 is the dielectric constant, m_0 is the free electron mass, $\Delta_{nn'}(k_{//}) \equiv |E_{n'}(k_{//}) - E_n(k_{//})|$, $E_n(k_{//})$ is the energy of subband n obtained by solving Eq. (2.19), $P_{nn'}(k_{//})$ is the momentum matrix element, $f_n(k_{//})$ is the carrier distribution function associated with subband n, and $\Gamma_{nn'}(k_{//})$ is the average scattering rate for the electronic states $|n, k_{//}\rangle$ and $|n', k_{//}\rangle$. The momentum matrix element $P_{nn'}(k_{//})$ for the inter-valence-band transitions can be derived from the $k \cdot p$ theory [61].

Chapter 3

3. P-DOPED GAINAS/ALGAAS STRAINED MQW STRUCTURES

3.1. SAMPLE GROWTH

Several p-type multi-quantum well (MQW) structures based on ternary compound semiconductors were grown. A buffer layer of 150nm was first grown followed by a Be-doped conduction layer of 1 μm. Then, the MQW structures were grown, which contained 35 periods of 3.0 nm wide $Ga_{0.85}In_{0.15}As$ wells separated by 30 nm wide undoped $Al_{0.33}Ga_{0.67}As$ barriers. Finally, a 0.3 μm of p^+ GaAs was grown as a cap layer. All the layers in the structure were grown at 600 °C except the $Ga_{0.85}In_{0.15}As$ wells which were grown at 530 °C to avoid the desorption of indium. A growth interruption time of 90 seconds was introduced between the quantum well layers to accommodate different growth temperatures. A (2 × 4) As-stabilized surface reconstruction was maintained throughout the growth of the whole structure. The wells were doped with Be of 2×10^{19} cm^{-3}, 1×10^{18} cm^{-3} and 1×10^{17} cm^{-3} for different samples. Another p-doped $Ga_{0.85}In_{0.15}As/Al_{0.45}Ga_{0.55}As$ sample was also grown. It has the same structure as that of $Ga_{0.85}In_{0.15}As/Al_{0.33}Ga_{0.67}As$ MQWs, but different Be concentration in the wells, different of Al composition and a cap layer of 0.5 μm.

The details of these MQW structures are summarized in Table 3.1. The column of Composition gives the mole fractions of Gallium in well and Aluminum in barrier. Data in the Doping densities show the doping type and doping concentration in the corresponding wells. The structure of all the MQW samples is illustrated in figure 3.1.

Table 3.1. Details of Samples

Structure (well / barrier)	Sample Lables	Composition(x)		Width (Å)		Periods	Doping Density (cm^{-3})
		Well (Ga)	Barrier (Al)	Well	Barrier		
Ga$_x$In$_{1-x}$As/Al$_x$Ga$_{1-x}$As	A	0.85	0.33	30	300	35	Be, 1×10^{17}
	B	0.85	0.33	30	300	35	Be, 1×10^{18}
	C	0.85	0.33	30	300	35	Be, 2×10^{19}

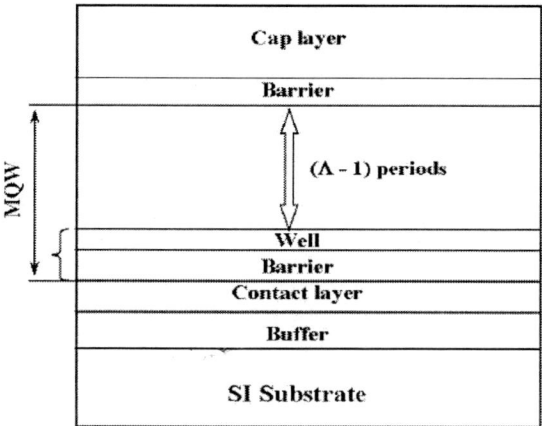

Figure 3.1. The Quantum Well Structure.

3.2. BAND OFFSET DETERMINATION

The energy band lineups of the semiconductor heterojunctions and superlattices are the essential and important parameters in the theoretical estimation of band structures. The Model-Solid Theory (MST) [62] was a simplified model to determine the band offset. The basic idea is to setup an absolute reference energy level and all calculated energies can then be put on an absolute energy scale. In the model-solid theory, an average energy over the three uppermost valence bands (the heavy-hole, the light-hole, and the sin-orbit split-off bands) $E_{v,av}$ is obtained from theory and is referred to as the absolute energy level. The MST provides a simple guideline for estimating the band offsets for the materials, especially the ternary compounds with varying compositions for which experimental data may not be always available.

For unstrained semiconductors, the heavy hole and light hole band edges (E_{HH} and E_{LH}) are degenerate at the zone center, and their energy position is denoted as E_v,

$$E_v \equiv E_{v,av} + \frac{\Delta}{3} \qquad (3.1)$$

where Δ is the spin-orbit splitting energy, and the spin-orbit split-off band-edge energy E_{SO} is

$$E_{SO} = E_v - \Delta = E_{v,av} - \frac{2\Delta}{3} \qquad (3.2)$$

The conduction band edge is obtained by adding the band-gap energy E_g to E_v,

$$E_c = E_v + E_g \qquad (3.3)$$

The band lineups between material A and B are shown in figure 3.2. The band-edge discontinuities are

$$\Delta E_c = E_c^A - E_c^B, \quad \Delta E_v = E_v^B - E_v^A \qquad (3.4)$$

$$\Delta E_c + \Delta E_v = \Delta E_g \qquad (3.5)$$

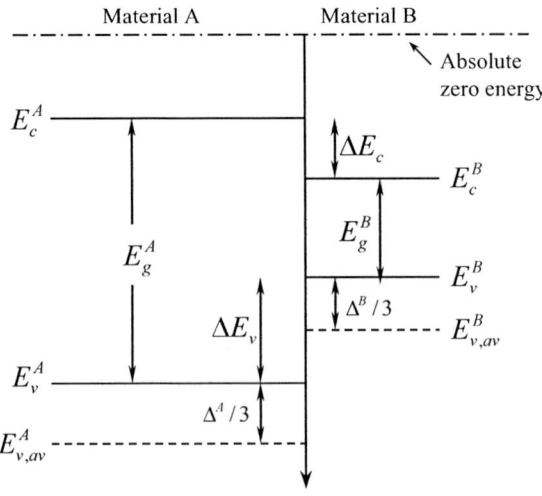

Figure 3.2. Band lineups in the heterostructures described by the MST.

The partition ratios of the band-edge discontinuities can be obtained,

$$Q_c = \Delta E_c / \Delta E_g, \quad Q_v = \Delta E_v / \Delta E_g \qquad (3.6)$$

If the material A with a lattice constant a is grown on a substrate with a lattice constant a_0 along z direction, we have

$$\varepsilon_{xx} = \varepsilon_{yy} = \frac{a_0 - a}{a} \text{ and } \varepsilon_{zz} = -2\frac{C_{12}}{C_{11}}\varepsilon_{xx} \qquad (3.7)$$

The band-edge shifts are

$$\Delta E_{v,av} = a_v(\varepsilon_{xx} + \varepsilon_{yy} + \varepsilon_{zz}) \equiv -P_\varepsilon \qquad (3.8)$$

$$\Delta E_c = a_c(\varepsilon_{xx} + \varepsilon_{yy} + \varepsilon_{zz}) \equiv -P_c \qquad (3.9)$$

The position of the average energy of the valence bands $E_{v,av}$ under strain is shifted from its unstrained position $E^0_{v,av}$ in Eq. (3.1) by $-P_\varepsilon$,

$$E_{v,av} = E^0_{v,av} - P_\varepsilon \qquad (3.10)$$

The center of the valence-band-edge energy is

$$E_{v,av} = E_{v,av} + \frac{\Delta}{3} = E^0_v - P_\varepsilon \qquad (3.11)$$

The heavy hole, light hole, and spin-orbit split-off band edges are

$$E_{HH} = E^0_v - P_\varepsilon - Q_\varepsilon \qquad (3.12)$$

$$E_{LH} = E^0_v - P_\varepsilon - \frac{\Delta}{2} + \frac{Q_\varepsilon}{2} + \frac{1}{2}\left[\Delta^2 + 2\Delta Q_\varepsilon + 9Q_\varepsilon^2\right]^{1/2} \qquad (3.13)$$

$$E_{SO} = E^0_v - P_\varepsilon - \frac{\Delta}{2} + \frac{Q_\varepsilon}{2} - \frac{1}{2}\left[\Delta^2 + 2\Delta Q_\varepsilon + 9Q_\varepsilon^2\right]^{1/2} \qquad (3.14)$$

The conduction band edge shifted by P_c is given by Eq. (3.9),

$$E_c = E^0_v + E_g(x) + P_c \qquad (3.15)$$

In the limit of a large spin-orbit split-off energy $\Delta \gg |Q_\varepsilon|$, we can ignore the coupling of the spin-orbit split-off band and

$$E_{LH} \cong E_v^0 - P_\varepsilon + Q_\varepsilon \quad (3.16)$$

$$E_{SO} \cong E_v^0 - P_\varepsilon - \Delta \quad (3.17)$$

For a ternary alloy such as $A_xB_{1-x}C$ with a lattice constant $a(x)$,

$$a(x) = xa(AC) + (1-x)a(BC) \quad (3.18)$$

which is a linear interpolation of the lattice constants of $a(AC)$ and $a(BC)$ of the binary compound semiconductors, the following formula is used to calculate an energy level E (e.g. $E_{v,av}^0$),

$$E(A_xB_{1-x}C) = xE(AC) + (1-x)E(BC) + 3x(1-x)[-a_v(AC) + a_v(BC)]\frac{\Delta a}{a_0} \quad (3.19)$$

where the last term accounts for a strain contribution to the ternary alloy, and $\Delta a = a(AC) - a(BC)$ is the difference between the lattice constants of two compounds AC and BC. Once $E_{v,av}^0$ is determined, the band-edge energies for the strained ternary compound can be calculated using Eq. (3.7) - (3.15).

As an application in $Ga_{0.85}In_{0.15}As/Al_{0.33}Ga_{0.67}As$ MQWs, the calculated values based on MST are tabulated in Table 3.2, the band gap energies, E_g of GaInAs and AlGaAs were taken from the experimental results [63,64] and the spin-orbit splitting Δ and lattice constant a were calculated by linearly interpolating the data obtained from the binary alloys [42]. Finally, the valence-band partition ratio Q_v of the MQWs is 39.2%, which is in good agreement with the experimental value of 40% [65].

Table 3.2. The calculated parameters based on MST

Parameters Materials	a (Å)	$E_{v,av}$ (eV)	Δ (eV)	P_ε (eV)	E_v (eV)	E_g (eV)
$Al_{0.33}Ga_{0.67}As$	5.656	-7.129	0.3202	0	-7.0213	1.8355
$Ga_{0.85}In_{0.15}As$	5.714	-6.878	0.346	0.0124	-6.775	1.208

3.3. PHOTOLUMINESCENCE MEASUREMENTS

3.3.1. Concentration Dependence of Band Gap

Figure 3.3 shows the 5 K Photoluminescence (PL) spectra of the three p-doped $Ga_{0.85}In_{0.15}As/Al_{0.33}Ga_{0.67}As$ MQW structures with Be-doping density of 1×10^{17} cm^{-3}, 1×10^{18} cm^{-3}, and 2×10^{19} cm^{-3} in the GaInAs well, respectively [66]. There are two luminescence peaks in each of the three samples. The two luminescence peaks for the sample with a doping concentration of 1×10^{17} cm^{-3} occurred at 1.492 eV and 1.572 eV, respectively. As the doping concentration in the well material increases to 1×10^{18} cm^{-3} and 2×10^{19} cm^{-3}, both peaks were found to shift to (1.488 eV and 1.565 eV) and (1.482 eV and 1.546 eV), respectively. In addition to the shift in the peak energy, the full width at half maximum (FWHM) of the luminescence peaks also varies. The FWHMs of the low energy peaks are 8 meV, 11.4 meV and 18.6 meV while that of the high energy peaks are 14.3 meV, 15.7 meV and 31.5 meV for the well doping densities of 1×10^{17} cm^{-3}, 1×10^{18} cm^{-3} and 2×10^{19} cm^{-3}, respectively. The wider PL linewidth at higher doping density implies more defects at interface and dispersions of the quantum energy levels due to more dopants.

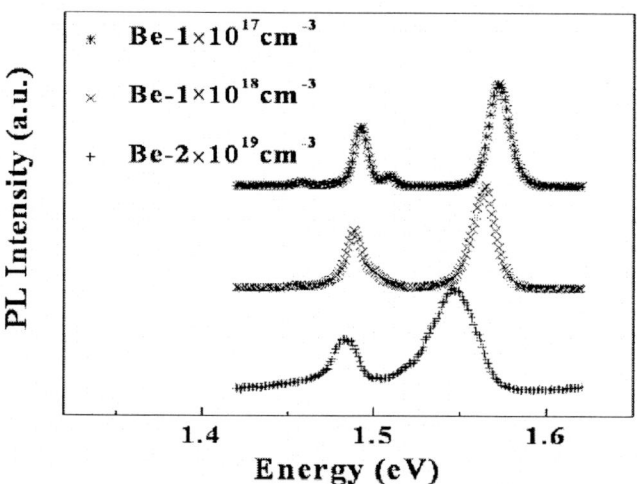

Figure 3.3. PL spectra of $In_{0.15}Ga_{0.85}As_{0.45}/Al_{0.33}Ga_{0.67}As$ MQW structures with different doping densities in the wells measured at 5 K, '*' - 10^{17} cm^{-3}, '×' - 10^{18} cm^{-3}, '+' - 2×10^{19} cm^{-3}.

To clarify the two peaks, the subband energy levels of the quantum well structures were calculated based on the envelope function approximation. The program is listed in Appendix A. It is well known that high doping causes the band-gap narrowing due mainly to the carrier interaction [67], and the concentration dependence of band-gap of the materials can be estimated based on the empirical equation given by Casey and Stern [68]. In our case, the band-gap offset in the valence band as a function of doping density can be expressed in the following form,

$$\Delta E_V = \Delta E_{V0} + Q_V (1.6 \times 10^{-8} p^{1/3}), \qquad (3.20)$$

where ΔE_{v0} and ΔE_v are the valence band offset of the QWs before and after taking doping-induced band gap narrowing into consideration. On the other hand, the biaxial strain is also a very important factor on the band structures. The valence band gap offset Q_v was taken as 40% of the total band gap offset as calculated in Section 3.3 and reported in Ref [65]. The $Ga_{0.85}In_{0.15}As$ and $Al_{0.33}Ga_{0.67}As$ parameters were linearly interpolated from the binary values and their energy gaps were taken from the reference [69]. The calculated band structures for the three samples are schematically illustrated in figure 3.4. Our estimation indicates that the higher energy peak is related to the transition between the ground state of electron C_1 in the conduction band and the ground state of light hole LH_1 in the valence band (C_1-LH_1), while the low energy peak to the transition between C_1 and the ground state of heavy hole HH_1 in the valence band (C_1-HH_1). As the doping density is increased, the PL peaks will shift towards the low energy side due to band-gap narrowing of the well materials. If the high energy PL peak for the structure with a doping density of 1×10^{17} cm^{-3} was taken as a reference, the shifts due to the band gap shrinkage with doping densities of 1×10^{18} cm^{-3} and 2×10^{19} cm^{-3} were 8.6 meV and 27 meV, respectively. They are in very good agreement with the red shifts of the C_1-LH_1 transition. The same conclusion can be drawn for the low energy peak resulted from the C_1-HH_1 transition. The increase in the FWHM with doping concentration can be explained by the spread of the subband energy levels in the wells. It indicates that high doping density not only enhances the barrier height mentioned earlier but also broadens the well-barrier interfaces.

In addition to HH_1 and LH_1, another subband energy level HH_2 is also expected in the well band structure, as indicated in figure 3.4. They all shift down with the increase in doping concentration. As a result, the energy difference between the subbands HH$_1$ and HH$_2$ is also increased due to the band gap narrowing of well material at high doping concentration.

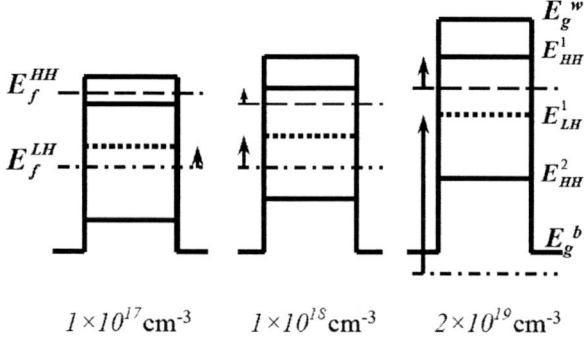

Figure 3.4. Subband energy and Fermi levels in the valence band of the wells for samples with different doping densities. Long dash – Fermi level for HH, dashed dot – Fermi level for LH, solid line – HH subbands, dot line– LH subbands.

The FWHM of the Photoluminescence line indicates the quality of the MQWs. The roughness, in terms of half-width distribution ΔL_z, resulted from doping can be estimated from the half width of the luminescence line using the following relation [70],

$$\Delta L_z = \Delta E_{pl} \left(\frac{\mu_0 L_z^3}{\hbar^2 \pi^2} \right) \tag{3.21}$$

where μ_0 is the reduced mass of exciton in the wells, ΔE_{pl} is the half width of the PL line and L_z is the (statistic) well width. The ΔL_z for the well of Sample A is 0.12 Å, indicating a high quality interface. By increasing the doping density to 10^{18} cm^{-3} and 2×10^{19} cm^{-3}, the half width ΔL_z becomes 0.18 Å and 0.4 Å, respectively. The latter is over three times that of the sample with lightly doped ones. The increasing half width due to the incorporation of Be dopant spreads the bound subbands of HH and LH and therefore increases the FWHM of the luminescence spectra of the QWIPs, originating from the interband recombination between the subband energy levels in the conduction band and the valence band. The increasing FWHM indicates the interface of the well-barrier becoming rough with the increasing doping density in the well.

3.3.2. PL Intensity and Linewidth at Various Temperatures

To further characterize the device structures, PL measurements were conducted at various temperatures. Figure 3.5 shows the PL spectra of the sample with a Be density of 2×10^{19} cm^{-3}. Both PL peaks shift towards long wavelength and their intensities decrease with the increase in temperature. As shown in figure 3.6(a), PL intensity for the two peaks follows an exponential relationship, expressed as [66]:

$$I_{PL}(T) = I_{PL}^c \exp(-T/T_0^c), \qquad (3.22)$$

where I_{PL}^c is a constant and T_0^c is the characteristic temperature. A larger T_0^c value implies less temperature-sensitive characteristics. The values of T_0^c are as large as 108 K and 165 K for the two peaks.

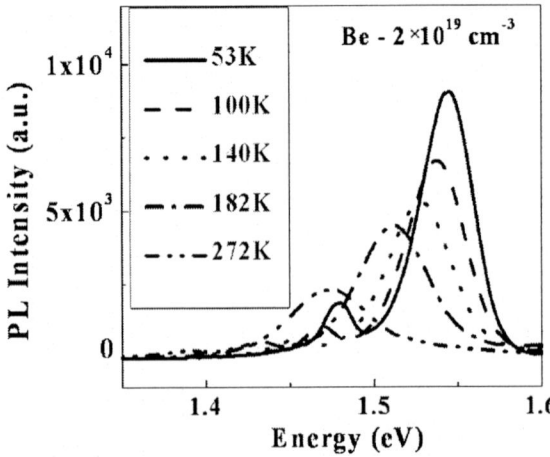

Figure 3.5. PL spectra of highly Be-doped In$_{0.15}$Ga$_{0.85}$As$_{0.45}$/Al$_{0.33}$Ga$_{0.67}$As MQW structures as a function of temperature. Solid line – 53 K, dash line – 100 K, dot line – 140 K, dash and dot line – 182 K, dash and dot-dot line – 272K.

The intensity ratios of C_1-LH_1 versus C_1-HH_1 transitions increase with the increase in temperature, as shown in figure 3.6(b). This increase is likely due to the increasing thermal population in the subband energy level LH_1 as temperature is increased. The FWHMs of the two PL peaks are shown in figure 3.7. The broadening of the PL peaks with temperature, as is usually observed, is due to thermal broadening.

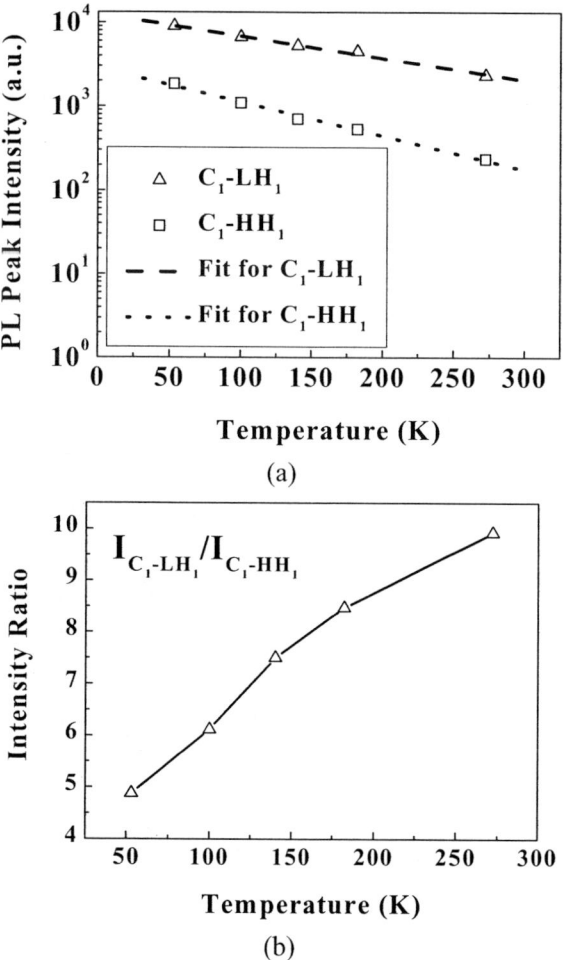

Figure 3.6. (a) Temperature dependence of PL peak intensities of C_1-HH_1 and C_1-LH_1 transitions of the $In_{0.15}Ga_{0.85}As_{0.45}/Al_{0.33}Ga_{0.67}As$ MQW structures with a Be density of 2×10^{19} cm^{-3} versus temperature. Open triangle – C_1-LH_1 transition, Open square –C_1-HH_1 transition. (b) Intensity ratio of C_1-LH_1 over C_1-HH_1.

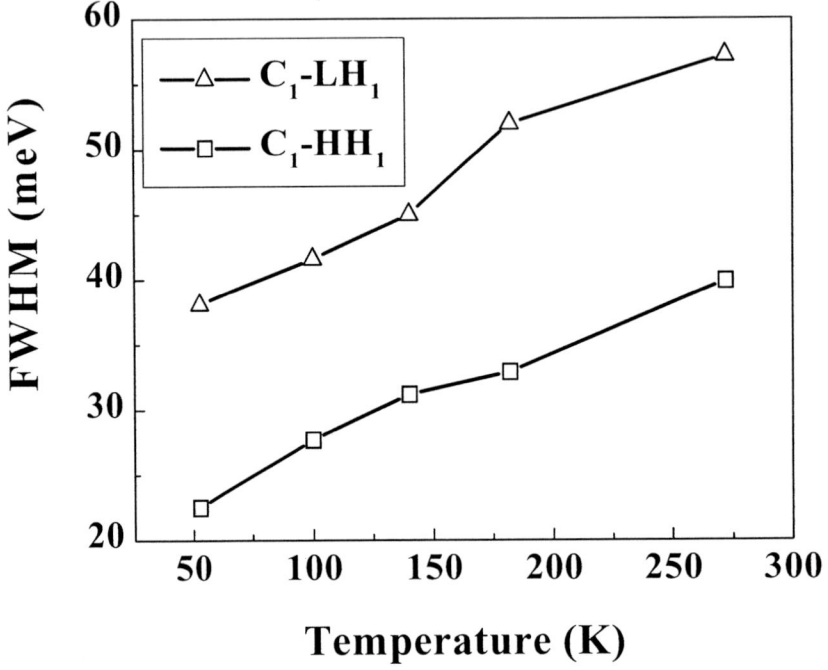

Figure 3.7. Full width at half maximum (FWHM) of the PL spectra of C_1-HH_1 and C_1-LH_1 transitions. Open triangle – C_1-LH_1, Open square – C_1-HH_1.

The energies of C_1-HH_1 and C_1-LH_1 transitions as a function of temperature are shown in figure 3.8. The Varshni's equation is also introduced for comparison [71],

$$E(T) = E(0) - \alpha T^2 /(\theta + T), \qquad (3.23)$$

where T is the absolute temperature in K, $E(0)$ is the band gap energy in eV at 0 K, and α and θ are fitting parameters. As indicated in the figure, the temperature dependence of the two luminescence peaks agrees well with the Varshni's relation. The α values are 5.4×10^{-4} and 5.3×10^{-4} and the θ values are 228.5 and 121.8, for the two peaks, respectively [66]. These values are slightly different from the data of bulk InGaAs [72]. This is understandable, as the two peaks in the GaInAs/AlGaAs MQWs are the C1-HH1 and C1-LH1 transitions in which the quantum size effect was also included.

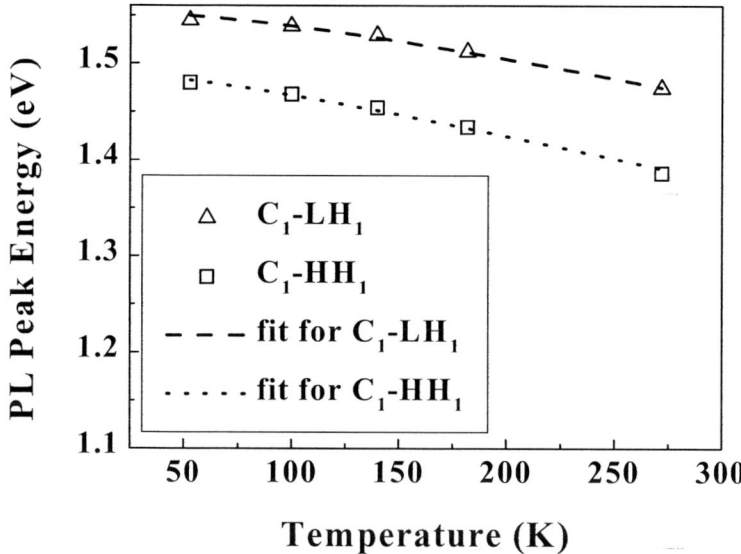

Figure 3.8. PL peak energies of C_1-HH_1 and C_1-LH_1 transitions of $In_{0.15}Ga_{0.85}As_{0.45}/Al_{0.33}Ga_{0.67}As$ MQW structure with a Be doping density of 2×10^{19} cm^{-3} versus temperature, and the fitting curves from Varshni's equation. Open triangle – C_1-LH_1, Open square – C_1-HH_1.

3.4. STRUCTURAL PROPERTIES

High-resolution x-ray diffraction (HRXRD) is a powerful tool to characterize semiconductor superlattices (SL) [73-76]. The information, such as the periodicity and mean mismatch, which are essential to the structure, can be obtained. Moreover the intensity and the line width of the diffraction Bragg peaks are good indicators of the quality of the MQWs, that is, the degree of the crystallinity and the uniformity of the layer thickness. The nominal periodicity of the MQW can be further confirmed by the cross-sectional high - resolution transmission electron micrograph (HRTEM).

3.4.1. Bragg Reflection Rocking Curves

Figure 3.9 shows the (004) GaAs Bragg reflection rocking curves plotted on a logarithmic scale using the 5 – crystal mode with the Bartels monochromator in the (220) setting and ω/2θ geometry [77]. The three group of spectra (a), (b) and (c) correspond to the three samples with doping density of 1×10^{17} cm^{-3}, 1×10^{18} cm^{-3} and 2×10^{19} cm^{-3} in the wells, respectively. The thicker solid lines are the results of experiments while the thinner lines correspond to the simulation results based on dynamical x-ray theory. For clarity, the spectra (b) and (c) have been shifted up with respect to the spectra (a). The scan step size is 0.0005 ° and the scan range is varied from about 2.5° to 5.5° for different samples. The peaks with the largest intensity in the curves are from the GaAs substrate, and the others are from the GaInAs/AlGaAs multiple quantum well structures. As seen from the figure, well-defined periodic satellite peaks up to ninth – order are observed, indicating good layer periodicity [78]. The sharpness of the peaks represents the abruptness of the interfaces of the strained MQWs. Any imperfection, relaxation, or compositional inhomogeneity would cause loss of phase coherence and eliminate the satellite peaks [79].

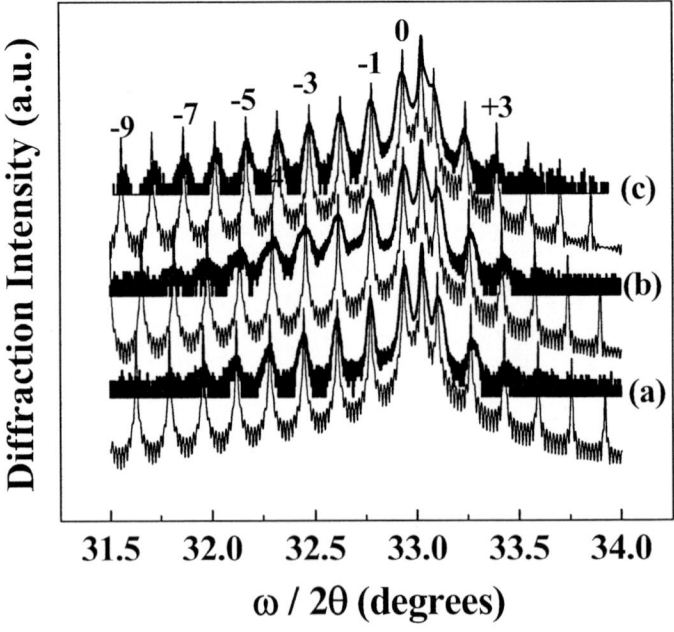

Figure 3.9. High resolution X-ray diffraction spectra for the three samples with doping density (a) 1×10^{17} cm^{-3}, (b) 1×10^{18} cm^{-3} and (c) 2×10^{19} cm^{-3} in the well. The dotted lines correspond to the simulation based on dynamical x-ray theory.

3.4.2. Average Mismatch

The angular separation $\Delta\theta_0$ between the zero-order satellite peak and GaAs substrate (004) reflection peak gives the average mismatch of the MQW along the growth direction,

$$\varepsilon \equiv \frac{\Delta a_\perp(\bar{x})}{a_0} = \frac{a_\perp(\bar{x}) - a_0}{a_0}, \qquad (3.24)$$

$$\varepsilon = -\frac{\Delta\theta_0}{\tan\theta_B}, \qquad (3.25)$$

where $a_\perp(\bar{x})$ is the average lattice parameter along the growth axis of the epitaxially strained MQW, a_0 is the lattice constant of the substrate, θ_B is the Bragg angle for the (004) reflection of the GaAs substrate. Equation (3.25) is obtained by simply differentiating Bragg's law and making the small angle approximation for $\Delta\theta_0$. Hence, based on the relation shown above, the mean mismatch of the MQW structures can be calculated. They are 1.14×10^{-3}, 1.37×10^{-3} and 1.57×10^{-3} for the strained structures with doping concentration of 1×10^{17} cm^{-3}, 1×10^{18} cm^{-3} and 2×10^{19} cm^{-3} in the wells, respectively. As shown in figure 3.10 by the open triangles, the increase of doping concentration leads to an increase in the average compressive mismatch.

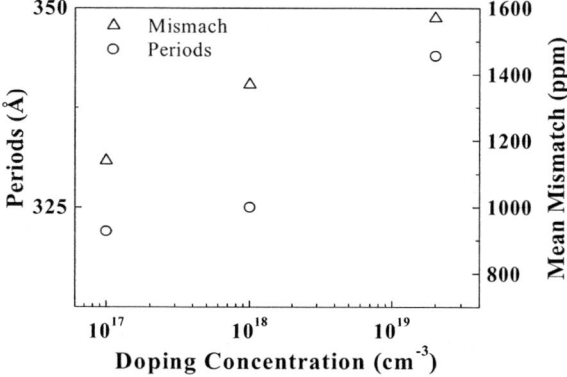

Figure 3.10. Periods and the mean mismatch for the three samples with different Be-doping density in the well. Open Triangle -Mean mismatch, Open Circle – Periods.

In general, the compressive strain is mainly caused by the incorporation of elements that have higher atomic number. If it is true, the high doping concentration of Be should not cause the increasing compressive strain if they are occupying the substitution site in the structures. We are not clear at this moment what is the mechanism behind the increase in the mismatch. As the atomic number of the Be is much smaller than that of the In and Ga group III elements, the Be is unlikely to increase the compressive strain by substitutionally occupying the sites of the group III elements. However, it is possible that as the doping density is increased, Be maybe incorporated into the well material interstitially. They may make the material tighter and cause additional strain in the tight lattices of the well material. Hence, we suspect that the increasing interstitial Be dopants at high doping concentration may play a role for the increasing mismatch.

3.4.3. Period of MQWs

The angular spacing $\Delta\theta$ between the satellite peaks is a measure of the MQW period length that is the sum of the InGaAs well and AlGaAs barrier layers thickness. The period Λ of the MQW structures is straightforwardly determined from the $\Delta\theta$ through the relationship [80],

$$\Lambda = \frac{\lambda}{2\Delta\theta \cos(\theta_B + \Delta\theta_0)}, \qquad (3.26)$$

where $\lambda = 1.54051$ Å for the Cu Kα_1 line, θ_B and $\Delta\theta_0$ are 33.03° for GaAs substrate Bragg peak and the angular separation between the substrate and the zero-order satellite, respectively. The periods extracted from Equation (3.26) were 322 Å, 326 Å and 337 Å for sample A, B and C, respectively. They also increase with the doping concentration in the wells of the MQW structures, as shown in figure 3.10 by the open circles. The possible reasons will be explained later in the Sec. 3.5.6.

3.4.4. Line-Width of the Zero-Order Peak

The full width at half maximum (FWHM) is often quoted as a figure of merit for a crystal. As for the superlattice structure, the FWHM of the zero-order is a parameter that represents the quality of the multilayer structure. Figure 3.11 show the FWHM of the zero-order satellite versus the Be-doping level. The FWHM is fitted using a Rectangular Hyperbola function for the range of doping levels studied as:

$$FWHM = P\frac{qD}{1+qD}, \qquad (3.27)$$

where D is the doping level, P and q are constants with values of 0.03534 ° and 2.2514×10^{-17} cm^3, respectively.

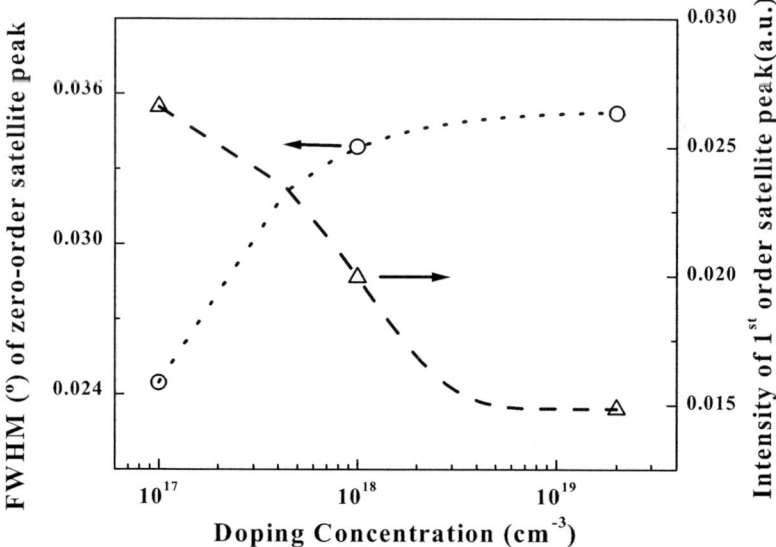

Figure 3.11. Full Width at Half Maximum (FWHM) of the zero-order satellite peak (…) and peak intensity of the first order satellite peak (---) v.s. doping concentration in the well.

The increasing FWHM indicates that the peak of the zero-order satellite is broadening with the increasing doping density. The peak broadening may not be due to spectral dispersion from the intrinsic line width of the characteristic x-ray line as it is completely eliminated by the Bartels monochromator. The increasing FWHM with the doping density further indicates that the interface imperfection and/or impurity diffusion and defects maybe enhanced by the higher doping density in the well materials.

3.4.5. Intensity of the First Order Peak

The intensity of the first order satellite peak, which is the characteristic of superlattice rocking curves, is strongly dependent upon diffusion [81]. The diffusion decreases the contact between the layers, and thus results in reduced satellite peak intensities particularly for the higher order satellites. The right Y-axis in figure 3.11 shows the intensity values of first order satellite peaks for three samples, which were normalized to that of GaAs substrate. They follow the exponential decay profile with the doping density in the following form:

$$I = Ae^{-D/D_0} + I_0, \qquad (3.28)$$

where D is the doping density and D_0 is a doping related constant with a vaule of 1.08×10^{18} cm^{-3}, A and I_0 are constants of 0.013 and 0.015, respectively. Lower value of D_0 indicates the higher doping effects on the lattice geometry structure. The decreased intensity of the first order satellite peak with the increased doping concentration above indicates the probable larger diffusion depth at higher doping density. It is understandable that the lattice defects such as interstitials due to high doping that was also described in Sec. 3.4.2 may enhance the diffusion processes.

3.4.6. Simulation Results

The rocking curves were simulated using the Philips simulation program HRS that was based on the solution of the Takagi-Taupin equations of the dynamical x-ray theory, as shown in figure 3.9. The thickness values of the barrier (well), 296 Å (25 Å), 302 Å (26 Å) and 315 Å (28 Å) for the three samples A, B and C, respectively, were used for matching the rocking curves in simulation. Compared to simulation, the overall decrease of the intensities of the experimental spectra is likely due to the presence of diffusion at the interfaces between the GaInAs and AlGaAs. The broadening of higher order satellite peaks is the effect of random lateral and the depth fluctuations of the period. One reason for the increase of barrier thickness in the simulation and the increase of the periodicity from rocking curves described in Sec. 3.4.3, may be due to the Beryllium doping which might change the quality of the GaInAs well layers, and thus affect the growth of the subsequent AlGaAs barrier layers. Gray et al. [82] have reported that HRXRD will not yield a unique solution for the width of the well and the barrier but rather a set of possible solutions due to the effective existence of an additional image barrier. However, the entire active MQW thickness obtained from HRXRD can be used. The values of the periods obtained from the simulation are 321 Å, 328 Å and 343 Å for sample A, B and C, respectively. These values are in good agreement with the results obtained by Eq. (3.26) based on the angular spacing between the satellite peaks that are 322 Å, 326 Å and 337 Å, respectively.

3.4.7. Transition Electron Microscopy

To verify the width of each individual layer, cross-section Transition Electron Microscopy (TEM) measurements were conducted. Due to the long speciman preparation time and the lack of abruptness at the heterojuction in cross sectional TEM, and at best the interface can be abrupt to within one lattice plane, only the sample C with the highest doping density was prepared and measured. Figure 3.12a shows the TEM image of the MQW structure, taken at the magnification of 20 K. The narrow dark line and the bright band orrespond to GaInAs wells and AlGaAs barriers, respectively. The high resolution TEM image was taken at the magnification of one million as shown in figure 3.12b. The well width and barrier width obtained from HRTEM are around 31 Å and 308 Å. The corresponding period of 339 Å is in good agreement with 337 Å from the XRD experiment and 343 Å obtained from the simulation as discussed in Section 3.5.6. The blur parts of the interface between GaInAs well and AlGaAs barrier in the figure 3.12b gave the evidence of the interdiffussion.

The uniform periodic structures are seen in the micrographs and the well thickness is close to the nominal 30 Å. However, the well width obtained by simulation in Sec 3.5.6 is smaller than the value measured from transmission electron microscopy and the nominal value expected from the growth rates and the known growth times. One reason may probably be due to the limitation of the HRS simulation package, as it cannot count the periodic fluctuation and the complicated interface situation. One has to consider the influence of interface nonplanarity and the roughness on the satellite intensities as well. Yashar et. al simulated their x-ray spectrum of the InGaAs/GaAs superlattice and reported the best fittings by taking the composition and layer thickness modulation into account [83]. But, the trend in this case, such as the increasing periods with doping density, indicated that the higher doping concentration in the well might have affected the growth condition for the barrier. From the simulation of the experimental spectra, the increasing of the well width with the increased doping density also concide with the increased mean strain obtained from HRXRD.

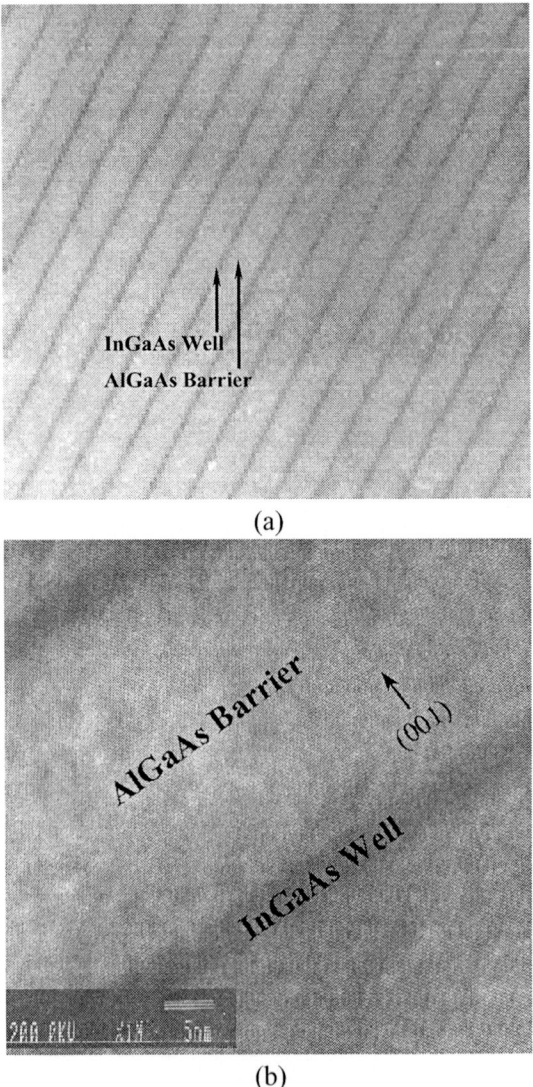

Figure 3.12. High resolution transmission electron micrography of the sample at doping density of 2×10^{19} cm^{-3} in the well with a magnification of (a) 20 K and (b) 1 Million.

3.5. INTERSUBBAND ABSORPTION OF THE P-TYPE GAINAS/ALGAAS MULTIPLE QUANTUM WELLS

3.5.1. Theoretical Approach

The recent theoretical approach for strained layers based on Luttinger-Kohn Hamiltonian has been demonstrated by solving the 2×2 matrix, including the mixings of heavy hole and light hole bands [84]. Furthermore, the coupling between the holes bands and spin-orbit split-off bands was taken into account [85] and compared to the results of the 2×2 matrix in InGaAs/InGaAsP and InGaAs/InP quantum well structures. However, by far, to our best knowledge, there is little report on numerical calculation related to the p-well-doping effects on MQWs. In the following part, the theoretical estimation of Be-doping effects on the dispersion of energy level using the six-band $k \cdot p$ method is discussed..

3.5.2. Six-Band $k \cdot p$ Model and Transfer Matrix Method

For the p-doped case, only the valence bands are of interest. Based on the theory of Luttinger-Kohn, the six-valence subbands i.e. the doubly degenerated heavy-hole, light-hole, and spin-orbit split-off bands are included. The valence-band structure of a strained bulk semiconductor under the cubic approximation can be described by a 6×6 Luttinger-Kohn Hamiltonian including the effect of strain as Equations (2.11) – (2.14). The eigen energies $E_v^i(k_{//}^{v,i})$ and the eigen wave functions $F_v^i(k_{//}^{v,i}, z^i)$ of the valence subband in the ith layer for the MQW material can be obtained by solving the effective-mass equation,

$$\sum_\mu (H_{\mu\nu}^i - V^i(z^i)) F_\mu^i(k_{//}^{v,i}, z^i) = E_v^i(k_{//}^{v,i}) F_v^i(k_{//}^{v,i}, z^i), \quad (3.29)$$

$$V^i(z) = \begin{cases} -\Delta E_v & in \ barrier \\ 0 & in \ well \end{cases}, \quad (3.30)$$

$$\mu,\nu \in \left\{ \left|\frac{3}{2},\frac{3}{2}\right\rangle, \left|\frac{3}{2},\frac{1}{2}\right\rangle, \left|\frac{1}{2},\frac{1}{2}\right\rangle, \left|\frac{3}{2},-\frac{3}{2}\right\rangle, \left|\frac{3}{2},-\frac{1}{2}\right\rangle, \left|\frac{1}{2},-\frac{1}{2}\right\rangle \right\}, \quad (3.31)$$

where $H_{\mu\nu}^i$ is the Luttinger-Kohn Hamiltonian in Eq. (2.11), $F_\nu^i(k_{//}^i, z^i)$ is the wave function which can be written as a linear combination of the plane waves in each region, $k_{//}^{v,i} = k_x^{v,i}\hat{x} + k_y^{v,i}\hat{y}$ is the in-plane wave vector, $V^i(z)$ is the barrier height of the QW, and ΔE_v is the valence band offset of the quantum well structure.

By introducing the continuities of the envelope functions and the probability current density at the boundary between the well and the barrier, the subband energies and the linear combination coefficient of the envelope function can be worked out. Considering the ith well layer sandwiched by the $(i-1)^{th}$ and $(i+1)^{th}$ barrier layers in a MQW structure and assuming that the first layer is the barrier material and the value of i can only be an even integer. Thus, the entire envelope wave function F_v^i in the ith layer which includes the incident $\vec{F}_v^i$ and the reflected wave function $\overline{F}_v^i$ can be expressed as:

$$F_v^i = \begin{pmatrix} \vec{F}_v^i \\ \overline{F}_v^i \end{pmatrix}. \quad (3.32)$$

The two interfaces between the well and the barrier are located at:

$$Z_1^i = (i/2 - 1)L + L_b, \text{ and } Z_2^i = (i/2)L \quad (3.33)$$

where L is the period of the QW and L_b is the barrier width. The transmission matrix T_i for the ith layer can be expressed as a 12×12 matrix,

$$T_i = \begin{bmatrix} T_i^{11} & T_i^{12} \\ T_i^{21} & T_i^{22} \end{bmatrix}, \quad (3.34)$$

where

$$T_i^{11} = [e^{jk_z^{v,i} z^i} \delta_{\mu,v}], \qquad (3.34a)$$

$$T_i^{12} = [e^{-jk_z^{v,i} z^i} \delta_{\mu,v}], \qquad (3.34b)$$

$$T_i^{21} = T_i' \cdot T_i^{11}, \qquad (3.34c)$$

$$T_i^{22} = T_i' \cdot T_i^{12}, \qquad (3.34d)$$

$$T_i' = \begin{bmatrix} (\gamma_1^i - 2\gamma_2^i)\frac{\partial}{\partial z} & -j\sqrt{3}\gamma_3^i k_{//}^{v,i} & -j\sqrt{\frac{3}{2}}\gamma_3^i k_{//}^{v,i} & 0 & 0 & 0 \\ j\sqrt{3}\gamma_3^i k_{//}^{v,i} & (\gamma_1^i + 2\gamma_2^i)\frac{\partial}{\partial z} & 0 & 0 & 0 & j\frac{3}{\sqrt{2}}\gamma_3^i k_{//}^{v,i} \\ j\sqrt{\frac{3}{2}}\gamma_3^i k_{//}^{v,i} & 0 & \gamma_1^i \frac{\partial}{\partial z} & 0 & j\frac{3}{\sqrt{2}}\gamma_3^i k_{//}^{v,i} & 0 \\ 0 & 0 & 0 & (\gamma_1^i - 2\gamma_2^i)\frac{\partial}{\partial z} & -j\sqrt{3}\gamma_3^i k_{//}^{v,i} & j\sqrt{\frac{3}{2}}\gamma_3^i k_{//}^{v,i} \\ 0 & 0 & -j\frac{3}{\sqrt{2}}\gamma_3^i k_{//}^{v,i} & j\sqrt{3}\gamma_3^i k_{//}^{v,i} & (\gamma_1^i + 2\gamma_2^i)\frac{\partial}{\partial z} & 0 \\ 0 & -j\frac{3}{\sqrt{2}}\gamma_3^i k_{//}^{v,i} & 0 & -j\sqrt{\frac{3}{2}}\gamma_3^i k_{//}^{v,i} & 0 & \gamma_1^i \frac{\partial}{\partial z} \end{bmatrix} \qquad (3.34e)$$

where $k_z^{v,i}$ represents the wave vectors in the z direction for the vth state in the ith layers and is interpreted as a differential operator $-i\frac{\partial}{\partial z}$. γ_1^i, γ_2^i and γ_3^i are the Luttinger inverse mass parameters in the ith layers.

Due to the continuities of the wave function and probability current at the interface z_1^i between the $(i-1)^{th}$ barrier and the ith well and z_2^i between the ith well and $(i+1)^{th}$ barrier, respectively, the transmission equation can be expressed as:

$$T_{i-1}\big|_{z_1^i} F_v^{i-1} = T_i\big|_{z_1^i} F_v^i \qquad (3.35)$$

$$T_i\big|_{z_2^i} F_v^i = T_{i+1}\big|_{z_2^i} F_v^{i+1} \qquad (3.36)$$

The final 12x12 transfer matrix T_{tran} for the quantum well is

$$T_{tran} = (T_{i+1}|_{Z_2^i})^{-1} \cdot T_i|_{Z_2^i} \cdot (T_i|_{Z_1^i})^{-1} \cdot T_{i-1}|_{Z_1^i} \qquad (3.37)$$

Because the wave function should reduce to zero at $\pm\infty$, there is no reflection in the last layer, and hence the transmission coefficient T_c^l for the HH, LH and SO can be expressed as:

$$T_c^l = \left| T_{tran}(l,l) - \frac{T_{tran}(l+6,l) \cdot T_{tran}(l,l+6)}{T_{tran}(l+6,l+6)} \right|^2, \qquad (3.38)$$

where l is an integer with a value of 1, 2 and 3, which represents the spin degenerate state of HH, LH and SO, respectively, $T_{tran}(l,l)$ is the (l,l) element in the T_{tran} transfer matrix..

3.5.3. Calculated Energy Levels and Intersubband Transition at Various Conditions

Our calculation was limited to the $Ga_{0.85}In_{0.15}As/Al_{0.33}Ga_{0.67}As$ MQW structures where the width of the well and the barrier was assumed to be 30 Å and 300 Å, respectively, and only the well material was doped by Be element. The main reason to select this structure was that there would be two heavy-hole subbands in the p-wells, which gives a long-wavelength infrared absorption. It can be verified by experiments resulting from the intersubband transitions between the two energy states of the heavy hole.

A. Constant Compressive Strain

The energy band structures of the $Ga_{0.85}In_{0.15}As/Al_{0.33}Ga_{0.67}As$ MQW structures were estimated based on the Luttinger-Kohn model using transfer matrix method introduced in the previous section. Here, the biaxial compressive strain approximation was used for simplicity:

$$\varepsilon_{xx} = \varepsilon_{yy} \neq \varepsilon_{zz},$$
$$\varepsilon_{xy} = \varepsilon_{yz} = \varepsilon_{zx} = 0. \qquad (3.39)$$

since ε_{ij} is the symmetric strain tensor, thus from Eq. (2.14)

$$R_\varepsilon = S_\varepsilon = 0,\qquad(3.40)$$

which can essentially be applied to the two most important strained systems, i.e. a strained-layer semiconductor pseudomorphically grown on a (001)-oriented substrate and a bulk semiconductor under an external uniaxial stress along the z direction. For the case of the lattice-mismatched strain, the symmetric strain tensors are

$$\varepsilon_{xx} = \varepsilon_{yy} = \frac{a_0 - a}{a},\qquad(3.41)$$

$$\varepsilon_{zz} = -\frac{2C_{12}}{C_{11}}\varepsilon_{xx},\qquad(3.42)$$

where a_0 and a are the lattice constants of the substrate and layer material, respectively and C_{11} and C_{12} are the elastic stiffness constants. When the biaxial compression strain approximation was applied to the $Ga_{0.85}In_{0.15}As/Al_{0.33}Ga_{0.67}As$ multiple quantum well structures, the valence band energy level dispersion along (001) direction with compressive strain effect on the $Ga_{0.85}In_{0.15}As$ bulk material is shown in figure 3.13. The valence band gap offset Q_V was assumed to be 40% of the total band gap offset as calculated in Sec 3.2. The binary material parameters used here were taken from the Ref.[42], and the $Ga_{0.85}In_{0.15}As$ and $Al_{0.33}Ga_{0.67}As$ parameters were linearly interpolated from the values of binary compositions except their energy gap values that were directly taken [42].

From figure 3.13 it is clearly seen that the compressive strain split the fourfold degenerate multiplet at the valence-band edge into a pair of degenerate doublets, which are of spin degeneracy. It pushes up the heavy-hole band by 21.3 meV whilst pulls down the light–hole band by 39.2 meV. Hence, the energy separation between the HH and LH is increasingly induced by the compressive strain. Based on Eq. (3.38), the transmission coefficients at $\vec{k} = 0$ in the valence band of the $Ga_{0.85}In_{0.15}As/Al_{0.33}Ga_{0.67}As$ QW were calculated and the results are shown in figure 3.14. It is seen that there are two resonant states for HH (solid line) and one resonant state for LH (dash line), indicating that there are two energy levels for the heavy hole and one energy level for the light hole.

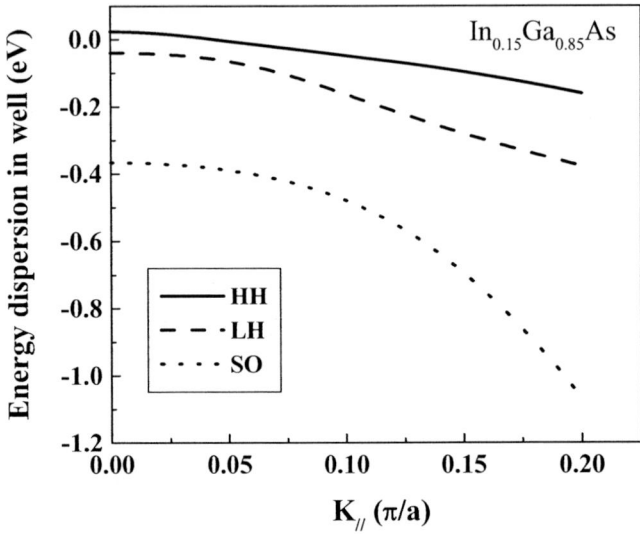

Figure 3.13. Calculated in-plane valence band energy dispersion of $In_{0.15}Ga_{0.85}As$ well material with the compressive strain effects. a is the lattice constant of well material in Å. Solid line – HH band, dash line – LH band, dot line – SO split off band.

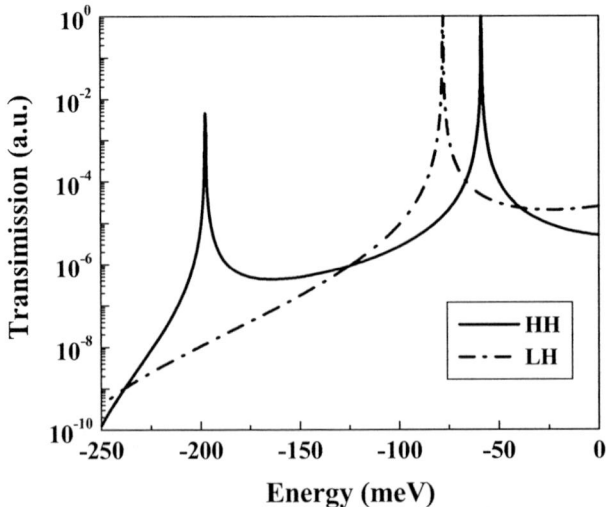

Figure 3.14. Calculated transmission coefficient with compressive strain taken into account for HH (Solid line) and LH (Dash dot line) near the Γ point for the $In_{0.15}Ga_{0.85}As/Al_{0.33}Ga_{0.67}As$ quantum well structures with 3 nm well width.

B. Doping Induced Bandgap Narrowing Plus Constant Compressive Strain

It is well known that high doping causes the band-gap narrowing, and hence it directly affects the barrier height of the quantum well as the well material is doped. Casey and Stern [68] gave an empirical equation to estimate the concentration dependence of the effective energy bandgap due to high doping,

$$E_g = E_{g0} - 1.6 \times 10^{-8} (p^{1/3} + n^{1/3}), \qquad (3.43)$$

where p and n are the hole and electron concentration, respectively, and the E_{g0} is the energy band gap in eV of the material with zero doping. In the case of p-type wells, the barrier height of the QW is not a constant but a function of doping density, which can be written in the following form, also shown in Eq. (3.20),

$$\Delta E_V = \Delta E_{V0} + Q_V (1.6 \times 10^{-8} p^{1/3} + n^{1/3}),$$

where ΔE_{V0}, ΔE_V are the valence band offset of the QWs before and after taking doping-induced band gap narrowing into consideration, respectively. By taking the band gap variation into account, together with the constant compressive strain introduced above, the subband energy levels of the holes in the valence band as a function of the doping density were shown in figure 3.15. It is seen that the subband energy levels move further away from the top valence band with the increasing doping density and the energy difference between the subbands of the heavy hole is also increased. These changes in the subband energy levels with the doping density are clearly due to the changes in the bandgap of the well material with doping. The increasing energy difference between the subbands implies that the corresponding wavelengths due to the intersubband absorption between them will blue shift with the increasing doping density.

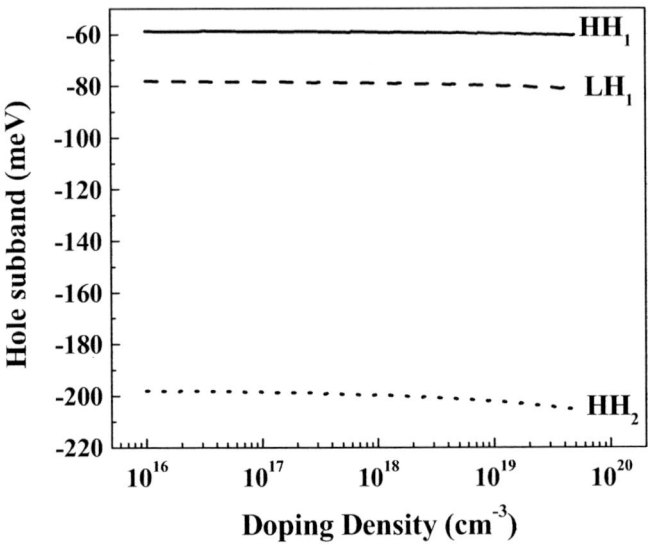

Figure 3.15. Calculated energy levels as a function of doping density for the ground states HH_1 (solid line) and LH_1 (dash line) and the first excited state HH_2 (dot line). Both compressive strain and doping density are taken into account for the $In_{0.15}Ga_{0.85}As/Al_{0.33}Ga_{0.67}As$ quantum well structures with 3 nm well width.

C. Strain and Barrier Variations with Doping

The calculated subband energies and their corresponding bound-to-bound transition wavelengths under different conditions and listed in Table 3.3. The calculated values with the compressive strain based on Eq. (2.14) and Eq. (3.41) taken into consideration are listed in the 'Constant strain' column. They are all the same for the three samples with different doping densities in the well layer as the strains based on the Eq. (3.41) have no relation with the doping density. By using the doping related empirical relation expressed in Eq. (3.20), the values of band gap shrinkage energy of the wells are estimated as 7.4 meV, 16 meV and 43.4 meV for the three samples A, B and C, respectively. They are listed in Table 3.4. The corresponding BTB transition wavelengths are 8.88 μm, 8.83 μm and 8.65 μm for the three samples A, B and C, respectively, as shown in Table 3.3.

Table 3.3. Heavy-hole subband energies and the corresponding intersubband transition wavelengths calculated under different considerations. Also included in the last column are the absorption wavelengths of the three samples with different doping in the well material, measured using Fourier transform infrared technique based on intersubband transition

Samples (Be-doping density cm^{-3})	HH sub-bands	Constant strain		Constant strain and bandgap shrinkage		Varied strain and bandgap shrinkage		Measured λ_p
		Sub-band energy	Transition λ	Sub-band energy	Transition λ	Sub-band energy	Transition λ	
		meV	μm	meV	μm	meV	μm	μm
A (1×10^{17})	HH$_1$	58.7	8.92	58.9	8.88	58.4	8.80	8.35
	HH$_2$	197.7		198.5		199.3		
B (1×10^{18})	HH$_1$	58.7	8.92	59.2	8.83	56.9	8.59	8.20
	HH$_2$	197.7		199.7		201.2		
C (2×10^{19})	HH$_1$	58.7	8.92	60.1	8.65	56.7	8.34	8.00
	HH$_2$	197.7		203.4		205.3		

Table 3.4. Bandgap energy and the compressive strain of the sample A, B and C, in which the Be doping concentration in the wells are 1×10^{17} cm^{-3}, 1×10^{18} cm^{-3}, 2×10^{19} cm^{-3}, respectively

Sample No.	Band gap Energy Shrinkage (meV)	Compressive Strain (x 10^{-3})
A	7.4	1.14
B	16	1.37
C	43.4	1.57

As mentioned above, the doping-induced band gap narrowing of the well material and a constant compressive strain have been considered for the calculation. In other words, the strain in the structures was assumed the same regardless of the change in the doping density. X-ray Diffraction (XRD) measurements for the three samples were conducted as described in Sec 3.5. The results are shown in figure 3.9. The angular separation between the zero-order satellite peak and GaAs substrate (004) reflection peak gives the average mismatch of the MQW along the growth direction. The mismatch values are 1.14×10^{-3}, 1.37×10^{-3} and 1.57×10^{-3} for the samples A, B and C, respectively, with the increasing doping density [86] as listed in Table 3.4 and discussed in Sec 3.4.2.

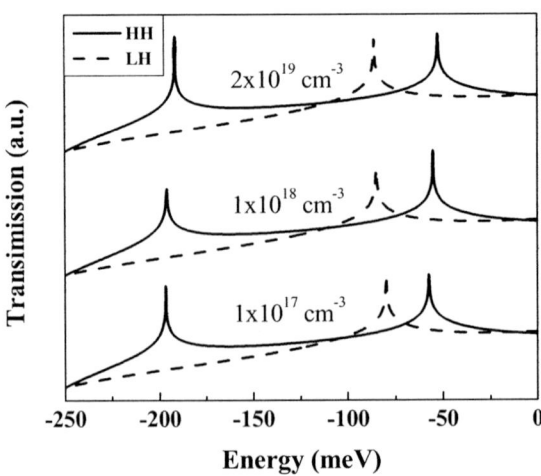

Figure 3.16. Calculated transmission as a function of energy for the three $In_{0.15}Ga_{0.85}As/Al_{0.33}Ga_{0.67}As$ MQW structures A, B and C with doping induced changes in strain and bandgap shrinkage taken into consideration. Solid line – HH, dash line – LH.

Based on the changes in the strain and the bandgap shrinkage caused by the doping density, the energy levels and the corresponding transmissions were recalculated. The results are shown in figure 3.16 and summarized in the 'Varied strain and bandgap shrinkage' column of Table 3.3. It is seen that the subband energy levels HH_1 and HH_2 and LH_1 vary for the three samples with different doping densities, and so do their intersubband transitions between the two energy levels of the heavy hole. As a result, the corresponding intersubband transition wavelength has blue-shifted to 8.8 µm, 8.59 µm and 8.34 µm, respectively [87].

3.5.4. Experimental Results and Comparison with the Calculated Values

To verify our calculation results, the infrared absorption measurement was carried out by a Perkin Elmer 2000 Fourier transformation infrared (FTIR) spectrophotometer at room temperature with samples polished into a 45° multipass waveguide geometry. A four-beam condenser was used to enhance the intensity of the incident source. The wavelengths of the absorption λ_p measured for the sample A, B and C are 8.35 µm, 8.20 µm, and 8.00 µm, respectively, as shown in figure 3.17 and also listed in the last column of Table 3.3. As shown in figure 3.17, only one absorption peak is observed even at a doping density of 2×10^{19} cm^{-3}, different from that observed for the n-type GaInAs/InAlAs [88]. There are three features in the spectra: (1) as the Be concentration is increased, the absorption spectrum blue shifts; (2) the ratio of $\Delta\lambda/\lambda$ of the three spectra is about $(16\pm1)\%$; (3) the FWHM value increases with the doping density, as illustrated in figure 3.18, the FWHM values of the PL peaks, which are obtained from Sec. 3.4.2, are shown in the figure for comparison. The blue shift of the intersubband absorption is as expected and was implemented in the calculation above. The value of $(16\pm1)\%$ is far less than 35%, required for a bound-to-continuum transition (BTC), and clearly indicates a bound-to-bound (BTB) intersubband transition [89].

The theoretical approaches to the FTIR under different considerations are clearly seen from the Table 3.3. When the effective band gap under high doping density was taken into account, the infrared absorption wavelength was estimated to blueshift with the increasing doping density. But the blue shift was not enough to satisfy the measurement values. The values of absorption wavelength shift for the sample B and C compared to the sample A estimated as 0.56% and 2.6% by considering only the concentration dependence of the effective band gap. They become 2.4% and 5.2% when the increasing strain is taken into account, and become much closer to the 1.8% ad 4.2% as measured. The estimated results by considering the increasing strain and the band gap shrinkage with the doping are much closer to the measured values. However, there are still some differences due likely to the complication of the transition in the p-type MQWs when only the two effects of high doping were considered in the calculation. Other Be-doping effects should also be taken into consideration for better results.

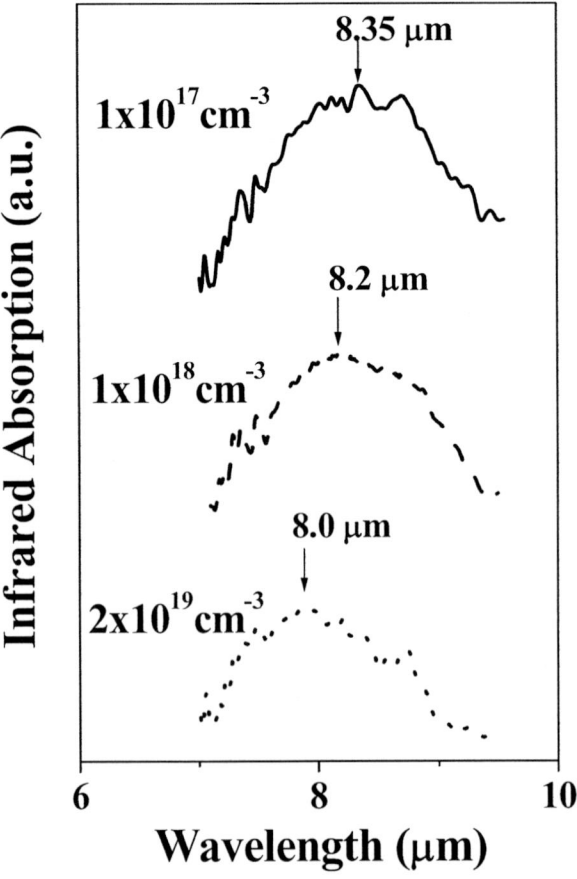

Figure 3.17. Measured FTIR spectra for the three samples with different Be doping densities.

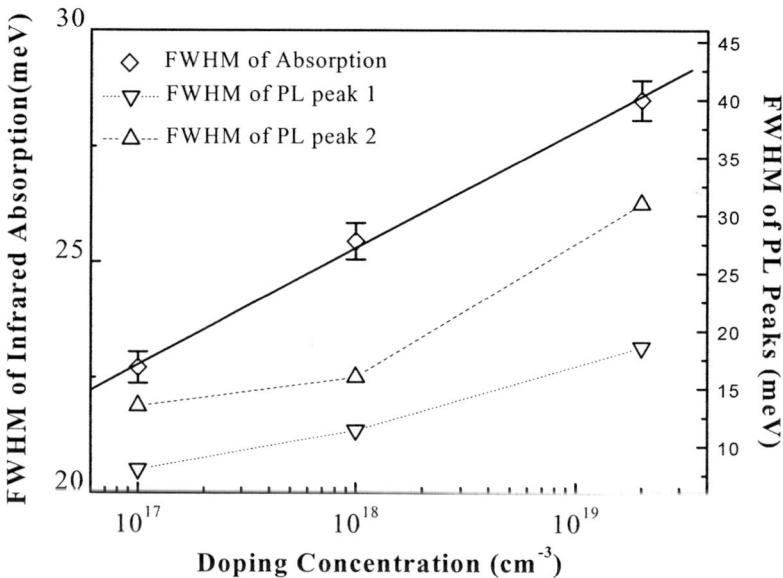

Figure 3.18. Plot of FWHM of the absorption spectrum as a function of doping density in the GaInAs well. The solid line is the linear fit. (The up and down triangles are the line width of the two PL peaks).

As shown in figure 3.18, the FWHM of the absorption spectra are about 23, 26, and 28 meV for the Sample A, B and C, respectively. The variation of doping density in the wells of the QWIPs with the FWHM fits an exponential relationship,

$$N_p = N_0 e^{(FWHM/C)} \qquad (3.44)$$

where N_p is the carrier concentration, and $N_0 = 7.88 \times 10^7$ cm^{-3}, and $C = 1.1$ meV are constants, and FWHM is in meV. It is worthwhile to point out that this relationship is similar to that of the doping density variation with the FWHM of the photoluminescence spectra of the Be-doped GaAs materials, but the two constants have different values [90]. The broadened FWHM of the absorption spectra with doping density indicates that the doping density deteriorates the well-barrier interface quality and enhances the impurity diffusion and defects. The results are consistent with the conclusions drawn from the photoluminescence line width in Sec 3.3.2 and the x-ray zero-order satellite peak in Sec 3.4.4.

Chapter 4

4. QUANTUM WELL INFRARED PHOTODETECTORS

4.1. DEVICE FABRICATIONS

After the structural, electrical and optical properties of the InGaAs/AlGaAs MQWs with different Be doping densities are investigated, we now look at the properties of devices. The samples were first etched into mesas of 200 μm in diameter by standard photolithography and a wet etching process. Gold-zinc was then evaporated on the p+ layers, followed by an annealing to form intimate ohmic contact. The device structures are shown in figure 4.1. The mesa top was taken as a positive terminal for electrical measurements.

The masks for the QWIP devices were fabricated by Taiwan Manufacture Corporation (TMC). Mask I is for the mesa island, Mask III and IV are for metal contact and Mask II is for gratings of n-type devices, including six kinds of cross gratings with different sizes, which are square cavities periodic in two orthogonal directions and both the top and the bottom of the cavities are covered with metal for light reflection. Their detection wavelength varied from 9μm to 14μm, tabulated in Table 4.1. λ_p is the center response wavelength, $\Lambda = a + b$ is the periodicity of the gratings. For the p-type Device structure, Mask I and Mask III are used. Figures 4.2 and 4.3 show the MQW structures after the photolithography process with above masks.

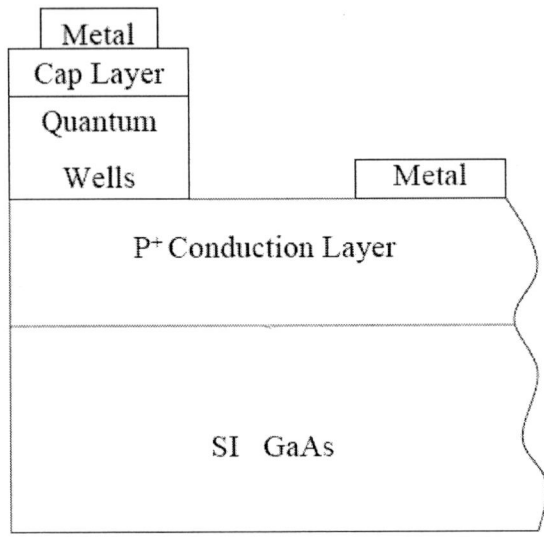

Figure 4.1. MQW device structure for dark current measurement.

Table 4.1. Details of the crossing gratings (Λ- periodicity)

λ_p (μm)	9	10	11	12	13	14
Λ (μm)	3.0	3.4	3.8	4.0	4.4	4.8
a×b(μm)	1.5×1.5	1.8×1.6	2.2×1.6	2.4×1.6	2.6×1.8	3.4×1.4

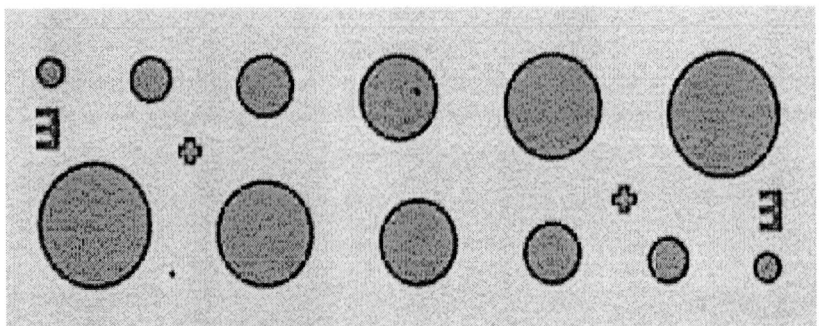

Figure 4.2. Mesa island. Round ones: mesas and cross: align symbol.

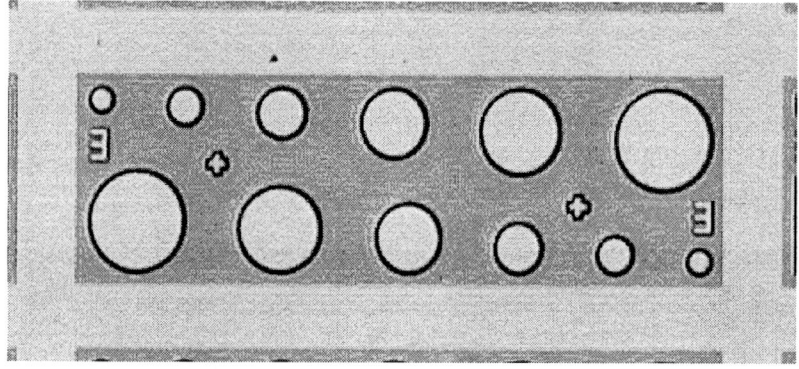

Figure 4.3. Metal contact (mainly used for p-type). Round and boundaries: metal contact, cross: align symbol.

Then the substrate near the devices is polished into a 45 degree angle to receive the incident light, as shown in figure 4.4.

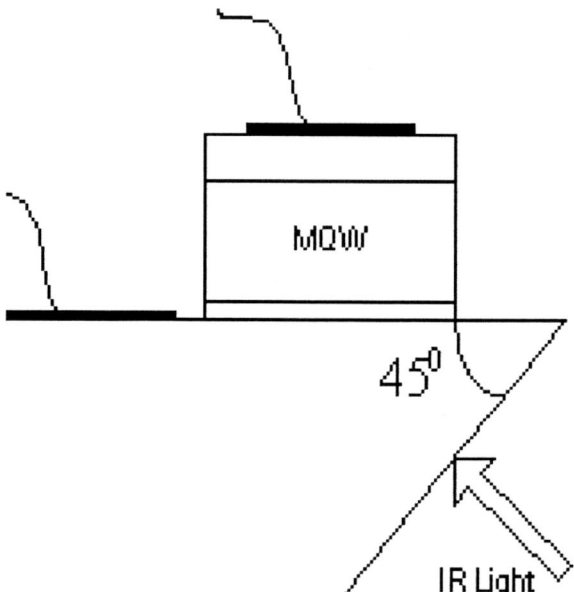

Figure 4.4 QWIP Device.

The full device fabrication process flow is shown in figure 4.5.

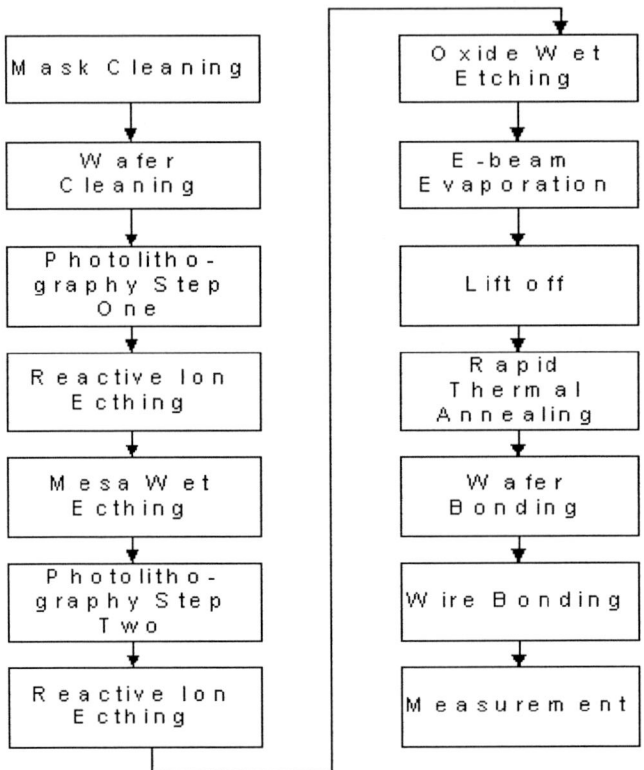

Figure 4.5. Device Processing flow.

4.2. DARK CURRENT

4.2.1. Background

The dark current consists of three components: thermionic emission, thermally assisted tunneling, and temperature-independent tunneling [91-94]. Thermionic emission refers to the carriers that are thermally excited into the current-carrying states above the top of the well; the resulting current varies exponentially with the temperature. The temperature-independent tunneling means tunneling of carriers into a neighboring well from densely occupied states at the "bottom" of the well. This current is only weakly temperature dependent, and it is observed at temperatures low enough to suppress thermionic emission.

Thermally assisted tunneling is the term usually used to refer to an intermediate case, in which carriers are thermally excited into states lying high in the confined band, but below the top of the well, and then tunnel through the triangular tip of the barrier (under an applied electric field) into the current-carrying states above the top of the barrier at a different location. This process is also called field emission, which is consistent with the fact that it is negligible compared to thermionic emission except at relatively high electric fields.

Equation (4.1) gives a very good account of both the temperature as well as the bias-voltage dependence of the dark current, $I_d(V)$ [92].

$$I_d(V) = n^*(V)ev(V)A, \qquad (4.1)$$

where $n^*(V)$ given by Equation (4.2) is the effective number of electrons which are thermally excited out of the well into the continuum transport states, e is the electronic charge, $v(V)$ is the average transport velocity given by Equation (4.3), and A is the device area.

$$n^*(V) = \left(\frac{m^*}{\pi\hbar^2 L_p}\right)\int_{E_1}^{\infty} f(E)T_{tf}(E,V)dE, \qquad (4.2)$$

where the first factor containing the effective mass m^* is obtained by dividing the two-dimensional density of states by the superlattice period L_p (to convert it into an average three-dimensional density). The Fermi factor $f(E)$ is given by

$$f(E) = \frac{1}{1 + e^{(E-E_1-E_F)/KT}}, \qquad (4.3)$$

where E_1 is the bound ground-state energy, E_F is the two-dimensional Fermi level (measured relative to E_1), and $T_{tf}(E,V)$ is the bias-dependent tunneling current transmission factor for a single barrier. The average transport velocity is given by

$$v(V) = \frac{\mu F}{\sqrt{1+(\mu F/v_s)^2}}, \qquad (4.4)$$

where μ is the mobility, F is the average field, and v_s is the saturated drift velocity.

A much simpler expression [95,96] which is a useful low-bias approximation can be obtained by setting $T_{tf}(E) = 0$ for $E < E_b$ and $T_{tf}(E) = 1$ for $E > E_b$, resulting in

$$n^*(V) = \frac{m^* KT}{\pi \hbar^2 L_p} e^{-(E_C - E_F)/KT}, \qquad (4.5)$$

where we have set the spectral cutoff energy $E_C = E_b - E_1$. Therefore,

$$\frac{I_d}{T} \propto e^{-(E_C - E_F)/KT}, \qquad (4.6)$$

where the Fermi energy level can be obtained from

$$N_D = n_0 \ln(1 + e^{E_F/KY}) \qquad (4.7)$$

and

$$n_0 \equiv \frac{m^* KT}{\pi \hbar^2 L_w}. \qquad (4.8)$$

4.2.2. Dark Current of P-Type Gainas/Algaas MQW Structures

The devices have been sampled and measured for the dark current. Figure 4.6 shows the measured dark current, I_d, of the p-type $Ga_{0.85}In_{0.15}As/Al_{0.45}Ga_{0.55}As$ strained MQW structures versus biased voltage, V_b (from −10 V to +10 V), as a function of temperature in the range of 5K - 150K. The current is found to be basically symmetric to the zero voltage and the value increased with the temperature T and V_b as expected. It is also found that the current in the structure is lower than that of the unstrained $GaAs/AlGaAs$ MQW structure. For example, at $V_b = 1V$ and at 80 K the current is about 1×10^{-9} A, about two orders of magnitude lower than that reported for the p-type $GaAs/AlGaAs$ QWIPs of the same size [97]. It is mainly due to the lower mobility of the heavier holes in $GaInAs$ well compared to that in $GaAs$ well. The small jumps at a current of about 10^{-9} A are due to software error that occurs when changing the current scale into log scale, which does not affect the data analysis.

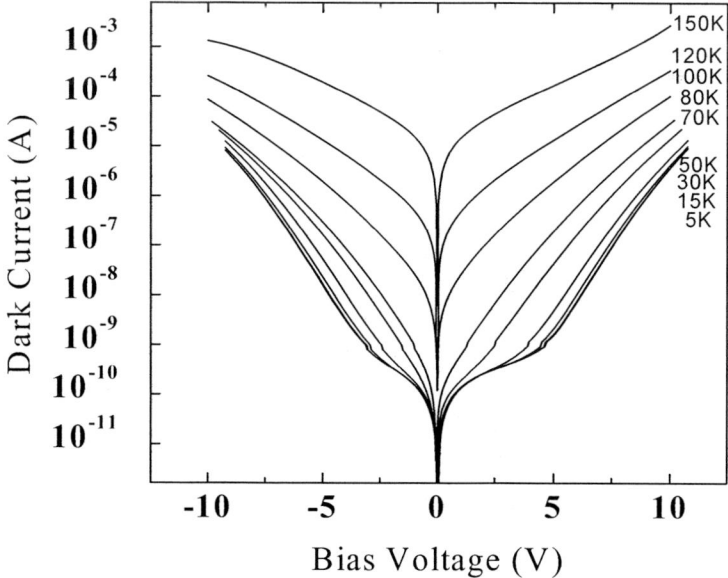

Figure 4.6. Dark current of the p-doped $Ga_{0.85}In_{0.15}As/Al_{0.45}Ga_{0.55}As$ MQW structure of 200 μm in diameter vs bias voltage at temperatures from 5 to 150 K.

The normalized dark current I_d/T versus $1000/T$ as a function of V_b is plotted in figure 4.7. It is clearly seen that over the biased voltage range, the normalized currents seem to fit two different expressions. At temperature of 70 K and above, it follows as

$$\frac{I_d}{T} \propto e^{-\frac{\Delta E}{KT}}, \qquad (4.9)$$

where the activation energy ΔE is the energy difference between E_c, the spectral cutoff energy, and the Fermi level E_F. This expression is the same as reported for the GaAs/AlGaAs material system [94]. However, at temperatures from 30K to about 5K, the dark current I_d did not change with temperature significantly but increased with V_b. This observation indicates that the dark current I_d involves two different transport mechanisms one of which dominates at one temperature range. A transition region for the two mechanisms lies between 70 K and 30 K. The thermionic emission dominates the conduction at temperature above 70 K, which is lower than 100 K for the n-type GaAs/AlGaAs MQWs and the tunneling mechanism dominates at temperatures below 30 K (figure 4.7). This discrepancy may be a result of the difference in the well depth of the different material systems. The MQWs studied here have a barrier height of 250 meV [81], which is much smaller than that for the unstrained GaAs/AlGaAs MQWs. The shallow well depth may need lower temperature for the thermionic emission to dominate.

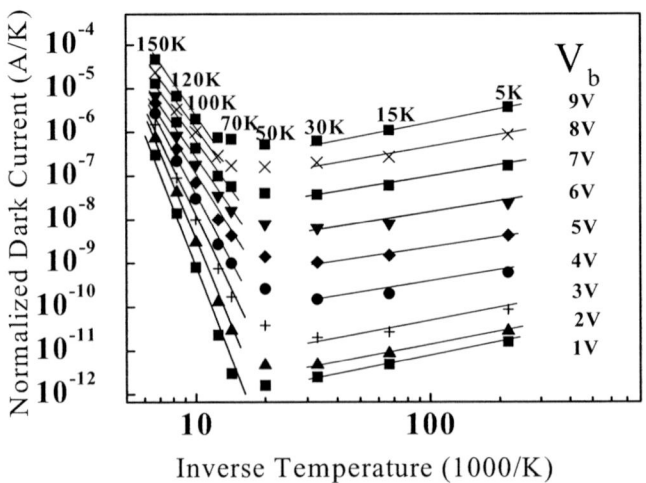

Figure 4.7. Plot of normalized dark current I_d/T vs $1000/T$ at different bias voltages, The lines are aid to the eye.

For the dark current at temperatures above 70 K, where the normalized current follows Equation (4.10), ΔE could be extracted from the current data and its relation with the bias voltage V_b is illustrated in figure 4.8. It is interesting to see that ΔE follows an exponential relationship with V_b [98],

$$\Delta E = Ce^{-\frac{V_b}{9}}, \qquad (4.11)$$

where C, extrapolated from the intercept on vertical axis, is equal to 156 meV. The exponential reduction of the activation energy ΔE with V_b (figure 4.8) is likely due to the energy bending.

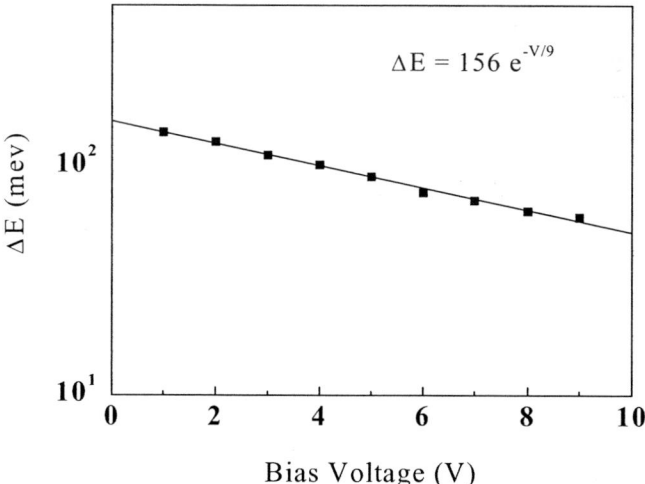

Figure 4.8. Plot of activation energy ΔE vs bias voltage.

There are two bound HH subbands, E_{hh1} and E_{hh2}, and one bound LH E_{lh1} inside the wells of the MQW structures based on our calculation described in section 3. Their energy levels measured from the top of the valence band are 41 meV, 160 meV and 99 meV, respectively, as shown in figure 4.9(a). At thermal equilibrium, the activation energy extrapolated from figure 4.8 is 156 meV. This value corresponds to a Fermi level located between HH_1 and LH_1 but closer to LH_1, or about 95 meV below the maximum of the valence band of the wells. As the bias voltage increases, the energy bands are not symmetrical any more and start to bend gradually. Thus the barrier over which holes are excited for conduction also becomes lower, as shown in figure 4.9(b). When the biased voltage is high enough, E_{hh2} may fall outside the well and the barrier becomes significantly low as shown in figure 4.9(c). The band bending due to biased field could qualitatively explain the drop in the activation energy ΔE.

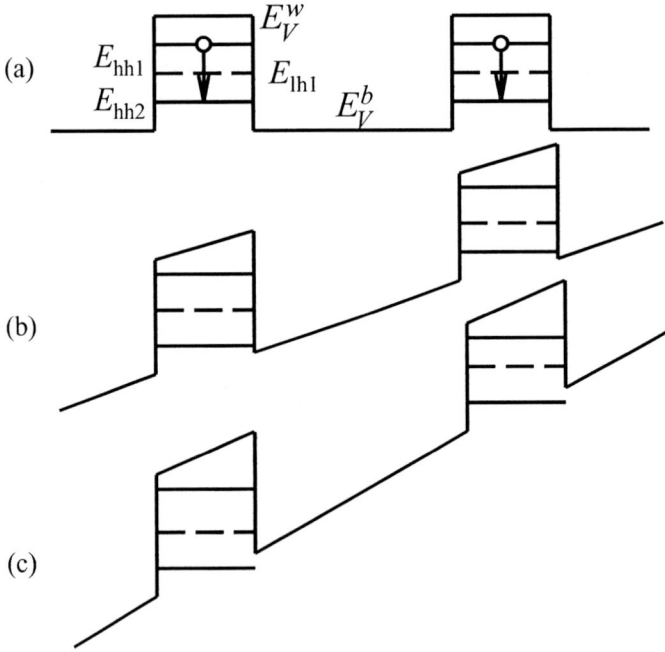

Figure 4.9. Schematic energy diagram of the MQWs (a) at thermal equilibrium, (b) at a moderate voltage, and (c) at a high voltage. E_{hh1} and E_{hh2} (solid lines) are two HH subbands, E_{lh1} (broken line) is the LH subband. E_V^w and E_V^b are the maximum of the valence band of the wells and barriers, respectively.

4.3. PERFORMANCE OF QWIP DEVICES

Two sets of p-type strained InGaAs/AlGaAs QWIPs, namely samples B and C, were measured here. They have the same structure parameters but different doping concentration in the wells, 10^{18} cm^{-3} and 2×10^{19} cm^{-3}, respectively. Figure 4.10 shows the photocurrent spectra of the two samples measured from the devices of 200 μm in diameter. Firstly, both devices show strong ptotoresponse. Secondly, the absorption peak wavelength, λp, of the sample B is about 6.2 μm while the sample C which has the same structural parameters but higher doping density in the wells is about 5.5 μm. Lastly, the ratios of width at half maximum $\Delta\lambda$ over the peak wavelength λp are about 27% and 20% for sample C and B, respectively, indicating a bound-to-continuum transition.

Although the two samples have the same structure parameters, the higher doping density in the wells will cause interface imperfection and bandgap narrowing. The reduced peak wavelength for the sample with higher doping density is likely due to the bandgap narrowing which causes the ground level of the heavy holes further away from the barrier edge and results in a shorter wavelength of absorption. It should also be pointed out that the photocurrent peak is at low wavelength compared to that measured using FTIR technique. This is because only the carriers, which can overcome the barriers after absorbing a photon, are able to contribute to the conduction current.

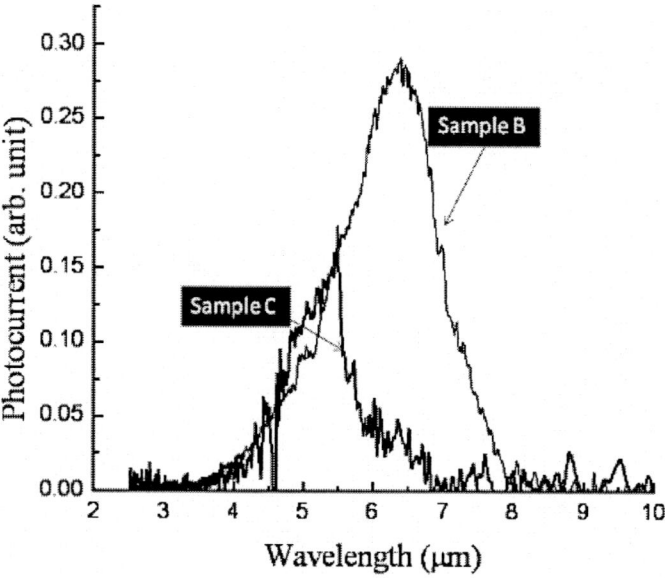

Figure 4.10. Photocurrent spectra of two QWIPs with a diameter of 200 μm. Both device consist of 35 InGsAs/AlGaAs quantum wells. The Be doping density in the wells are 10^{18} cm^{-3} (B) and 2×10^{19} cm^{-3} (C), respectively.

The responsivity of the sample with a Be concentration of 10^{18} cm^{-3} are shown as figure 4.11. Three frequencies of the incident light, controlled by a chopper, are employed during the measurements. The bias in the measurement is a current rather than a voltage as the resistance of the device is very high. As illustrated in the figure, the responsivity is the highest at a frequency of 600 Hz. It decreases as the frequency is increased to 800 Hz and 1000 Hz. This is mainly due to the time constant of the device circuit, with which the response signal at high frequency may not be able to follow the change of the incident. It is worth noting that the responsivity measured at two bias directions is basically symmetrical, which is advantageous compared to n-type and p-type QWIPs reported previously.

For the sample with the same structure parameters but different Be-doping in the wells, however, the symmetry of responsivity may be affected. Figure 4.12 shows the results from a sample with a doping concentration of 2×10^{19} cm^{-3} in the wells, which are also measured at three frequencies of chopped incident. It is found that the responsivity at forward bias has much higher values than that at reversed bias, and they are even higher than that of the sample with lower Be doping in the wells. The asymmetry of the resposivity is believed to be due to the inhomogeneity of the well-barrier junctions in the sample, as mentioned in the photocurrent results, while the higher responsivity values at forward bias is likely due to the enhanced carrier population from higher doping.

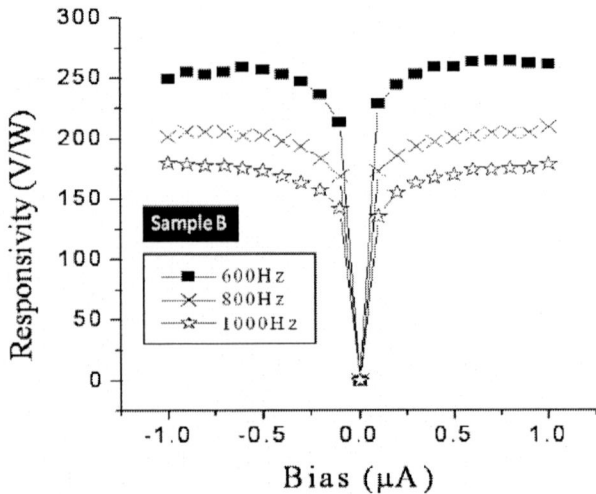

Figure 4.11. Responsivity of Sample B measured at three different frequencies.

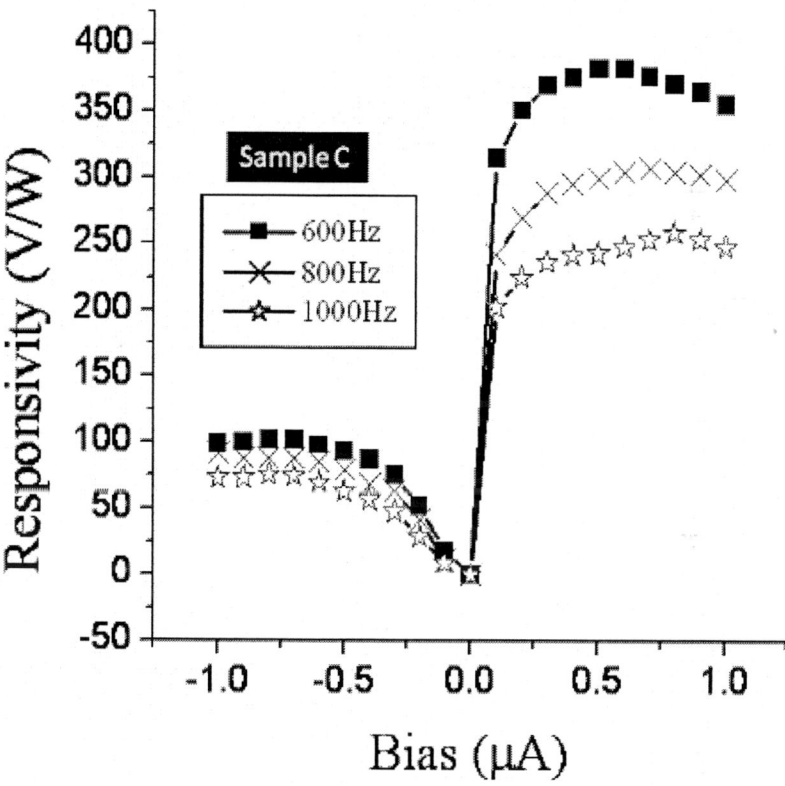

Figure 4.12. Responsivity of Sample C measured at three different frequencies.

The blackbody detectivities of the two QWIP devices with Be-doping concentrations of 10^{18} cm^{-3} and 2×10^{19} cm^{-3} in their wells, respectively, are shown in figure 4.13. The temperature of the blackbody was set at 500 K during measurements. As seen from the figure, detectivity of the sample B with 10^{18} cm^{-3} Be in the wells is symmetric while that of the sample with higher doping is asymmetric. It is understandable as the detectivity is proportional to the responsivity of the device. The detectivity of Sample B is about 8×10^{8} cm·Hz$^{1/2}$/W, which is comparable to and even better than that of the devices made up of other materials. For the device with higher Be doping in the wells, it can reach 10^{9} cm·Hz$^{1/2}$/W. It is indicated that the compressively strained InGaAs/AlGaAs QWIPs can be symmetric in forward and reversed biases as long as the doping concentration in the wells are not too high, and the detectivity can be further enhanced in one bias direction by high doping if the symmetry is not a concern in the application [99].

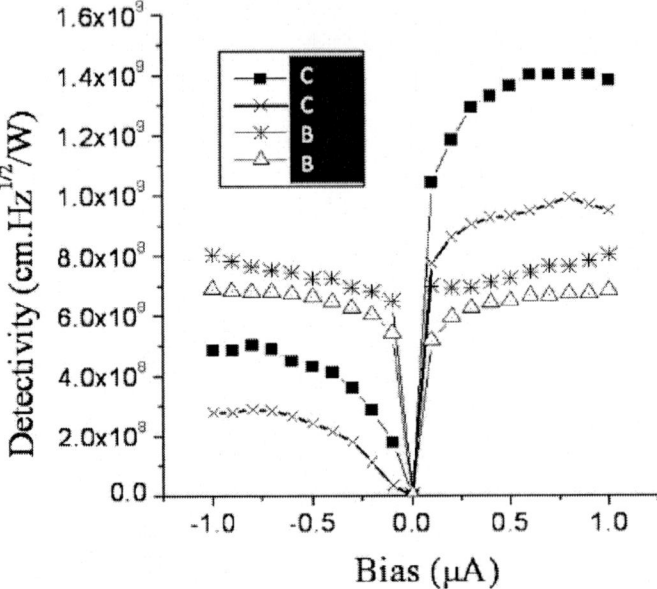

Figure 4.13. Detectivity of two samples, B and C, which have Be-doping concentration of 10^{18} cm^{-3} and 2×10^{19} cm^{-3} in the wells, respectively. They were measured at two frequencies of chopped incident (600 Hz and 800 Hz) with the blackbody set to 500 K.

Chapter 5

5. CONCLUSION

Effects of Be-doping density on optical, structural and Electrical properties of p-type GaInAs/AlGaAs multiple quantum-well (MQW) structures grown by SSMBE and performance of infrared photodetectors fabricated in house were extensively studied.

Two luminescence peaks were observed in the $Ga_{0.85}In_{0.15}As/Al_{0.33}Ga_{0.67}As$ strained MQW structures, one of which was related to C_1-HH_1 transition and the other to C_1-LH_1 transition. Both peaks shifted with the Be concentration in the well material of the structures. The experimental data were in good agreement with the theotical estimation based on the envelope function approximation, by taking the band-gap narrowing caused by Be doping and the biaxial compressive strain into consideration. The doping-caused difference in the subband energy levels HH_1 and HH_2, predicted by the theoretical estimation, was varified by the red shift of the photoluminescense line for the batch of samples with different doping densities in the wells. The increasing FWHM indicates that the quality of the well-barrier interfaces becomes worse with the increasing doping density in the well material. The intensities of the two PL peaks followed an exponential relationship with the temperature and the energies of the two PL peaks with the temperatures can be well described by the Varshni's equation.

The increasing average mismatch and period with the doping density were found by High-resolution X-ray diffraction spectra and TEM measurements. The FWHMs of the zero-order satellites fit a rectangular hyperbola function with the Be-doping level and the intensities of the first order satellite peaks for the three samples follow an exponential decay with doping density, indicating that the higher Be-doping density may enhance the Beryllium diffusion. The simulation of the X-ray spectra indicated that the Be-doping density in the well might cause different growth modes/rates for the consecutive barrier layer. The clear micrograph in high resolution obtained by TEM further confirmed the active region in the MQW stacks.

The theoretical approach of full six-band model using transfer matrix to the p-type strained $Ga_{0.85}In_{0.15}As/Al_{0.33}Ga_{0.67}As$ quantum well structures was implemented and the full energy band structures were calculated around the zone center. The high p-type doping density in the well material was found to increase the compressive strain. By taking both the varied barrier height and the varied strain caused by the p-type doping into account, the calculated subband energy levels of the heavy hole in the valence band were in good agreement with the measured results by FTIR based on the intersubband transition. This approach gives a better picture of the p-type doping effects for the GaInAs/AlGaAs quantum well structures and provides useful information for designing and fabricating the strained p-doped GaInAs/AlGaAs QWIPs.

For the p-type strained $Ga_{0.85}In_{0.15}As/Al_{0.45}Ga_{0.55}As$ QWIPs, the dark current was found much lower than that of the p-type GaAs/AlGaAs MQW structures. Two different mechanisms of conduction were found. They are tunneling and thermionic emission, each dominating at different temperature ranges. A transition region for the two mechanisms exists between 70 K and 30 K. The exponential activation energy relation with the bias voltage, $\Delta E = 156\exp(-V/9)$, was due likely to the energy band bending at high bias voltage.

The compressively strained InGaAs/AlGaAs QWIP devices with a Be concentration of 10^{18} cm^{-3} in the wells show a strong ptotoresponse peaked at 6.2 μm. Its responsivity and blackbody detectivity are symmetric for forward and reversed bias, and comparable to and even better than some p-type QWIPs made of other material systems. By increasing Be doping in the wells, the detected peak wavelength becomes smaller and the responsivity and detectivity become asymmetric due to the bandgap narrowing at high doping and inhomogeneity near the well-barrier interfaces.

Acknowlegements

Authors would like to thank Prof. Yoon Soon Fatt, Drs. Zhang Penghua and Dr. Zheng Haiqun for growing samples and helpful discussions and Dr. Tanakorn Osotchan for assistance in theoretical calculations.

REFERENCES

[1] Wilson, J., Hawkes, J.F.B., *Optoelectronics an Introduction,* Prentice-Hall: Englewood Cliffs, N.J., 1983.
[2] Capasso, F., *J. Vac. Sci. Technol. B*, 1983, vol. 1, 457.
[3] Capasso, F., *Science*, 1987, vol. 235, 172.
[4] Rosencher, E., Vinter, B., and Levine, B.F., *Intersubband Transition in Quantum Wells*, Plenum, NY, 1992, 9-14.
[5] Chahine M.T., *Proceedings of Innovative Long Wavelength Infrared Detector Workshop*, Pasadena, CA, 1990, April 24-26, 3.
[6] Levine B.F., Zussman A., Kuo J.M., and Jong J.De, *J. Appl. Phys.,* 1992, Vol. 71, 5130.
[7] Chiu L., Smith J.S., Margalit S. and Yariv A., Appl. Phys. Lett., 1983, vol. 43, 331.
[8] Smith J.S., Chiu L.C., Margalit S., Yariv A., and Cho A.Y., *J. Vac. Sci. Technol. B*, 1983, vol. 1, 376.
[9] Coon D. D., and Karunasiri R. P. G. , *Appl. Phys. Lett.,* 1984, vol. 45, 649.
[10] Goossen K. W., and Lyon S. A., *Appl. Phys. Lett,* 1985, vol. 47, 1257.
[11] Levine B. F., Choi K. K., Bethea C. G., Walker J., and Malik R. J., *Appl. Phys. Lett.,* 1987, vol. 50, 1092.
[12] Levine B. F., Bethea C. G., Choi K. K., Walker J., and Malik R. J., *J. Appl. Phys.,* 1988, vol. 64, 1591.
[13] Levine B. F., *J. Appl. Phys.,* 1993, vol. 74, R1.
[14] Wang Y. C. and Li S. S., *J. Appl. Phys.,* 1993,vol. 74, 2192.
[15] Chang Y. C., and James R. B., *Phys. Rev. B,* 1989vol. 39, 12672.
[16] Levine B. F., Gunapala S. D., Kuo J. M., Pei S. S., and Hui S., *Appl. Phys. Lett.,* 1991,vol. 59, 1864.

[17] Gunapala S. D., Levine B. F., Ritter D., Hamm R., and Panish M. B., *J. Appl. Phys.,* 1992, vol. 71, 2458.
[18] Levine B. F., Bethea C. G., Hasnain G., Shen V. O., Pelve E., Abbott R. R., and Hsieh S. J., *Appl. Phys. Lett.,* 1990, vol. 56, 851.
[19] Liu H. C., Steele A. G., Buchanan M., and Wasilewski Z. R., *J. Appl. Phys.,* 1993, vol. 73, 2029.
[20] Harris J. J., Cleff J. B., Beall R. B., Castagne J., Woodbridege K., and Roberts C., *J. Cryst. Growth,* 1991, vol. 111, 239.
[21] Y. H. Wang, Sheng S. Li, J. Chu, and Pin Ho, *Appl. Phys. Lett.,* vol. 64, 1994.
[22] Liu H. C., LI L., Buchanan M., Wasilewski Z. R., Brown G. J., Szmulowicz F., and Hegde S.M.., *J. Appl. Phys.,* 1998, vol. 83.
[23] Chang Y. C., and James R. B., *Phys. Rev. B,* 1989, vol. 39, 12672.
[24] Teng D., Lee C., and Eastman L. F., *J. Appl. Phys.,* 1992, vol. 72, 1539.
[25] Man P. P. and Pan D. S., *Appl. Phys. Lett.,* 1992, vol. 61, 2799-2801.
[26] Levine B. F., Gunapala S. D., Kuo J. M., Pei S. S., and Hui S., *Appl. Phys. Lett.,* 1991, vol. 59, 1864-1866.
[27] Liu H. C., Buchanan M., and Wasilewski Z. R., *Appl. Phys. Lett.,* 1998, vol. 72, 1682-1684.
[28] Singh J., *NATO ASI Ser. B,* 1991 vol. 253, 653.
[29] Osbourn G. C., *Superlat & Microstr.,* 1985, vol. 1, 223.
[30] Luryi S., Pearsall T. P., Temkin H., et al., *IEEE Electron Devices Lett.,* 1986, EDL-7, 104.
[31] Volk M., Lutegon S., Marschner T., Stolz W., Gobel E. O., Christianen P. C. M., and Maan J. C., *Phys. Rev. B,* 1995, vol. 52, 11096.
[32] *IEEE J. Quantum Electron.* QE-30, 1994, 348-618; *J. Crystal Growth,* 1989, vol. 164, 263-290.
[33] Chu J. and Li S. S., *IEEE J. Quantum Electron,* 1997, vol. 33, 1104-1113.
[34] Bardeen J., "An improved calculation of the energies of metallic Li and Na," *J. Chem. Phys.,* 1938, vol. 6, 367-371.
[35] Seitz F., *The mordern Theroy of Solids,* McGraw Hill, New York, 1940, 352.
[36] Shockley W., "Energy band structures in semiconductors", *Phys. Rev.,* 1950, vol. 78, 173-174.
[37] Dresselhaus G., Kip A. F., and Kittel C., "Cyclotron resonance of electrons and holes in silicon and germanium crystals," *Phys. Rev.,* 1995, vol. 98, 368-384.

[38] Kane E. O., "Band structure of indium antimonide," *J. Phys. Chem. Solids,* 1957,vol. 1, 249-261.
[39] Kane E. O., "The k.p method," *Semiconductors and Semimetals*, vol. 1, Academic, New York, 1966.
[40] Luttinger J. M. and Kohn W., "Motion of electrons and holes in perturbed periodic fields," *Phys. Rev.,* 1955,vol. 97, 869-883.
[41] Luttinger J. M., "Quantum theory of cyclotron resonance in semiconductors: General theory," *Phys. Rev.*, 1956,vol. 102, 1030-1041.
[42] Chuang S. L., *Physics of Optoelectronic Devices*, Wiley, New York, 1995, 658-661, 709-710,
[43] West L. C. and Eglash S. J., *Appl. Phys. Lett.*, 1985, vol. 47, 1257.
[44] Zaluzny M., *Thin Solid Films,* 1981, vol. 76, 307.
[45] Ahn D. and Chuang S. L., *J. Appl. Phys.,* 1987,vol. 62, 3052.
[46] Hasnain G., Levine B. F., Bethea C. G., Logan R. A., Walker J., and Malik R. J., *Appl. Phys. Lett.,* 1989, vol. 54, 2515
[47] Sarusi G., Levine B. F., Pearton S. J., Bandara S. V., and Liebenguth R. E., *Appl. Phys. Lett.,* 1994,vol. 64, 960.
[48] Chen C. J., Choi K. K., Tidrow M. Z., and Tsui D. C., *Appl. Phys. Lett.*, 1996, vol. 68, 1446.
[49] Hasnain G., Levine B. F., Bethea C. G., Logan R. A., Walker J., and Malik R. J., *Appl. Phys. Lett.*, 1989, vol. 54, 2515.
[50] Anderson J. Y., Lundqvist L., and Paska Z. F., *Appl. Phys. Lett.,* 1991,vol. 58, 2264.
[51] Anderson J. Y., Lundqvist L., *Appl. Phys. Lett.,* 1991,vol. 59, 857.
[52] Anderson J. Y., Lundqvist L., *J. Appl. Phys.,* 1992,vol. 71, 3600.
[53] Chang Y. C. and Schulman J. N., *Appl. Phys. Lett.,* 1983,vol. 43, 536.
[54] Chang Y. C. and Schulman J. N., *Phys. Rev. B,* 1985,vol. 31, 2069.
[55] Altarelli M., *Phys. Rev. B,* 1985,vol. 32, 5138.
[56] Sanders G. D. and Chang Y. C., *Phys. Rev. B,* 1985,vol. 31, 6892.
[57] Sanders G. D. and Chang Y. C., *Phys. Rev. B,* 1987,vol. 36, 4849.
[58] Kane E. O., *J. Phys. Chem. Solids,* 1956, vol. 1, 82.
[59] Chang Y. C., James R. B., *Phys. Rev. B,* 1989,vol. 39, 12672.
[60] James R. B. and Smith D. L., *Phys. Rev. Lett.*, vol. 43, 1495, 1979; *Phys. Rev. B,* 1980,vol. 21, 3502.
[61] Kane E. O., *J. Phys. Chem. Solids*, 1956,vol. 1, 82.
[62] Van de Walle C. G., "Band lineups and deformation potentials in the model-solid theory," *Phys. Rev. B,* 1989, vol. 39, 1871-1883.

[63] Hellwege K. H., Ed., *Landolt-Börnstein Numerical Data and Functional Relationships in Science ad Technology*, New Series, Group III, 17a, Springer, Berlin, 1982; Groups III-V 22a, Springer, Berlin, 1986.
[64] El Allai M., Sorensen C. B., Veje E., and Tidemand-Petersson P., "Experimental determination of the GaAS and $Ga_xAl_{1-x}As$ band-gap energy dependence on temperature and aluminum mole fraction in the direct band-gap region," *Phys. Rev. B*, 1993, vol. 48, 4398-4404.
[65] Radhakrishnan K., Yoon S. F., Li H. M., Han Z. Y., and Zhang D. H., "Photoluminescence studies of strained $In_xGa_{1-x}As/Al_{0.28}Ga_{0.72}As$ heterostructures grown by molecular-beam epitaxy", *J. Appl. Phys.*, 1994, vol. 76, 246-250.
[66] Shi W., Zhang D. H., Zhang P. H., Yoon S. F., "Optical properties of p-type InGaAs/AlGaAs multiple quantum well structures", *Microelectronic Engineering*, 2000, vol. 51-52, 181-187.
[67] Singh J., *Physics of Semiconductors and their heterostructures*, McGraw-Hill, 1993.
[68] Casey H. and Stern F., Concentration-dependent absorption and spontaneous emission of heavily doped GaAs, *J. Appl. Phys.*, 1976, vol. 47, 631-643.
[69] Madelung O., Semiconductors, *Group IV Elements and III-V Compounds*, Springer-Verlag, Berlin, 1991.
[70] Herman M. A., Bimberg D., and Christen J., *J. Appl. Phys.*, 1991, vol. 70, R1.
[71] Varshni Y. P., *Physica*, 1967, vol. 34, 149.
[72] Yu P. W., Kuphal E., *Solid State Commun.*, 1984, vol. 49, 907.
[73] Speriosu V. and Vreeland T., Jr., *J. Appl. Phys.*, 1984, vol. 56, 1591.
[74] Fleming R., Mcwhan D., Gossard A. C., Wiegmann W., and Logan R. A., J. Appl. Phys., 1980, vol. 51, 357.
[75] Vandenberg J., Hamm R. A., Macrander A. T., Panish M. B., and Temkin H., Appl. Phys. Lett., 1986, vol. 48, 1153.
[76] Macrander A. T., Schwartz G. P., and Gualtieri G. J., J. Appl. Phys., 1988, vol. 64, 6733,
[77] Shi W., Zhang D. H., Zhang P. H., Yoon S. F., "Optical properties of p-type InGaAs/AlGaAs multiple quantum well structures", Microelectronic Engineering, 2000, vol. 51-52, 181-187.
[78] Batterman B. and Hildebrandt G., Acta Crystallogy., Sect. A: Cryst. Phys. Diffr., Theor. Gen. Crystallogr., 1968, A24, 150.
[79] Tapfer L., Stolz W., and Ploog K., J. Appl. Phys., 1989, vol. 66, 3217.

[80] Holy V., Kubena J., and Ploog K., Phys. Status Solidi B, 1990, vol. 162, 347.
[81] Fatah J. M., Harrison P., Stirner T., Hogg J. H. C., and Hagston W. E., J. Appl. Phys., 1998, vol. 83, 4037-4041.
[82] Gray A. L., Newell T. C., Lester L. F., and Lee H., J. Appl. Phys., 1999, vol. 85, 7664-7670.
[83] Yashar P., Pillai M. R., Mirecki-Millunchick J., and Barnett S. A., J. Appl. Phys., 1998, vol. 83, 2010-2013.
[84] Klein W. H. and Roth L. M., "Deformation potential in germanium from optical absorption lines for exciton formation", *Phys. Rev. Lett.*, 1959, vol. 2, 334-335.
[85] Chao C. Y. P. and Chuang S. L., "Spin-orbit-coupling effects on the valence-band structure of strained semiconductor quantum wells", *Phys. Rev. B*, 1992,vol. 46, 4110-4122.
[86] Zhang D. H., Shi W., Zhang P. H. and Yoon S. F., "Doping effects on p-type InGaAs/AlGaAs quantum well structures for infrared photodetectors grown by molecular beam epitaxy", *Jpn. J. Appl. Phys.*, 1999,vol. 38, L360-L362.
[87] Shi W., Zhang D. H. and Osotchan T., "Six-band *kp* approach investigation of effects of doping on intersubband transition in strained InGaAs/AlGaAs quantum well structures", *IEEE J. Quantum Electronics,* , 2000,vol. 36, No. 7, 835-841.
[88] Asai H. and Kawamura Y., *Appl. Phys. Lett.*, 1990,vol. 56, 1427.
[89] Zhang D. H., Shi W., Zhang P. H., Yoon S. F. and Shi X., "Effect of Be-doping on the absorption of InGaAs/AlGaAs strained quantum well infrared photodetectors grown by molecular beam epitaxy", *Appl. Phys. Lett.*, 1999,vol. 74, 1570-1572.
[90] Zhang D. H., Radhakrishnan K., and Yoon S. F., *J. Cryst. Growth*, 1994,vol. 48, 35.
[91] Choi K. K., Levine B. F., Malik R. J., Walker J., and Bethea C. G., *Phys. Rev. B.*, 1987,vol. 35, 4172.
[92] Levine B. F., Bethea C. G., Hasnain G., Shen V. O., Pelve E., Abbott R. R., and Hsieh S. J., *Appl. Phys. Lett.*, 1990,vol. 56, 851.
[93] Andrews S. R., and Miller B. A., *J. Appl. Phys.*, 1991,vol. 70, 993.
[94] Williams G. M., DeWames R. E., Farley C. W., and Anderson R. J., *Appl. Phys. Lett.*, 1992,vol. 60, 1324.
[95] Kinch M. A. and Yariv A., *Appl. Phys. Lett.*, vol. 55, 2093, 1989.
[96] Gunapala S. D., Levine B. F., Pfeiffer L., and West K., *J. Appl. Phys.*, 1991,vol. 69, 6517.

[97] Levine B. F., Bethea C. G., Glogovsky G., Stayt J. W., and Leibenguth R. E., *Semicond. Sci. Technol.*, 1991,vol. 6, C114.
[98] Zhang D. H., and Shi W., *Appl. Phys. Lett.*, 1998,vol. 73, 1095 - 1098.
[99] Zhang D. H., Shi W., Li N., Chu J.H., *J Appl Phys*, 2002,vol. 92 , 6287-6290.

INDEX

A

absorption, ix, 1, 2, 3, 11, 12, 13, 14, 38, 41, 43, 45, 47, 59, 70, 71
absorption coefficient, 4, 14
absorption spectra, 13, 47
activation, 56, 57, 64
activation energy, 56, 57, 64
aid, 56
alloys, 20
aluminum, 15, 70
annealing, 49
ash, 40
ASI, 68
asymmetry, 60

B

back, 12
band gap, 1, 9, 20, 22, 23, 26, 39, 41, 42, 44, 45
bandgap, ix, 1, 2, 41, 43, 44, 45, 59, 64
barrier, ix, 6, 15, 16, 22, 24, 30, 32, 33, 36, 37, 38, 41, 47, 53, 56, 58, 59, 60, 63, 64
barriers, 3, 5, 11, 15, 33, 58, 59
bending, 57, 58, 64
bias, ix, 8, 9, 53, 54, 55, 56, 57, 60, 61, 64
biaxial, 22, 38, 39, 63

blueshift, 45
boundary conditions, 6
broadband, 3
BTC, 45
buffer, 15

C

candidates, 1
carrier, 3, 11, 13, 14, 22, 47, 60
cavities, 49
chemical vapor deposition, 2
CO_2, 2
coherence, 28
components, 4, 52
composition, 15, 33
compositional inhomogeneity, 28
compound semiconductors, 15, 20
compounds, 17, 20
concentration, ix, 15, 21, 22, 29, 30, 31, 32, 33, 41, 44, 45, 46, 47, 58, 60, 61, 62, 63, 64
conduction, ix, 1, 5, 7, 8, 9, 11, 15, 18, 19, 22, 23, 56, 58, 59, 64
Congress, vi
coupling, 3, 8, 12, 13, 20, 35, 71
cross-sectional, 27
crystallinity, 27
crystals, 68
cyclotron, 69

D

data analysis, 55
decay, 32, 63
defects, 21, 31, 32, 47
deformation, 11, 69
degenerate, 9, 17, 38, 39
density, ix, 2, 11, 21, 22, 23, 24, 25, 27, 28, 29, 30, 31, 32, 33, 34, 36, 41, 42, 43, 44, 45, 47, 53, 59, 63, 64
deposition, 2
desorption, 15
detection, 1, 2, 3, 49
dielectric constant, 14
diffraction, 13, 27
diffusion, 31, 32, 47, 63
diffusion process, 32
dispersion, 31, 35, 39, 40
distribution, 14, 23
distribution function, 14
dopant, 23
dopants, 21, 30
doped, ix, 3, 11, 13, 15, 21, 23, 24, 35, 38, 41, 47, 55, 64, 70
doping, ix, 4, 15, 21, 22, 23, 27, 28, 29, 30, 31, 32, 33, 34, 35, 41, 42, 43, 44, 45, 46, 47, 49, 58, 59, 60, 61, 62, 63, 64, 71

E

earth, 2
eigenenergy, 7
electric field, 3, 8, 12, 53
electron, 1, 3, 5, 10, 14, 22, 27, 33, 34, 41
electron microscopy, 33
electrons, 1, 3, 5, 53, 68, 69
emission, 1, 52, 53, 56, 64, 70
energy, ix, 1, 2, 4, 5, 6, 11, 14, 17, 18, 19, 20, 21, 22, 23, 25, 26, 35, 38, 39, 40, 41, 42, 43, 44, 45, 53, 54, 56, 57, 58, 63, 64, 70
epitaxy, 70

equilibrium, 57, 58
ESO, 17
estimating, 17
etching, 49
exciton, 23, 71
eye, 56

F

fabricate, 2
fabrication, 2, 12, 51
Fermi, 23, 53, 54, 56, 57
Fermi energy, 54
Fermi level, 23, 53, 56, 57
flow, 51, 52
fluctuations, 32
Fourier, 43, 45
Fourier transformation, 45
FTIR, 45, 46, 59, 64
FTIR technique, 59
Full Width at Half Maximum, 31
FWHM, 21, 22, 23, 26, 30, 31, 45, 47, 63

G

GaAs, ix, 2, 3, 4, 15, 27, 29, 30, 32, 33, 44, 47, 55, 56, 64, 70
generation, 1
germanium, 68, 71
gratings, 12, 13, 49, 50
groups, 4
growth, 2, 3, 13, 14, 15, 29, 32, 33, 44, 64
growth modes, 64
growth rate, 33
growth temperature, 15
growth time, 33

H

Hamiltonian, 9, 35, 36
hardness, 2
heat, 1

Index

height, ix, 6, 22, 36, 41, 56, 64
heterojunctions, 17
heterostructures, 4, 18, 70
high resolution, 33, 64
high-speed, 2
HRS, 32, 33
HRTEM, 27, 33

I

imaging, 2
incidence, 3, 12
indicators, 27
indium, 15, 69
infrared, ix, 1, 2, 3, 12, 38, 43, 45, 63, 71
infrared light, 12
inhomogeneity, ix, 60, 64
injury, vi
InP, 3, 35
integrated circuits, 2
integration, 2
interaction, 9, 22
interface, 6, 21, 23, 31, 33, 37, 47, 59
interstitial, 30
interstitials, 32
intrinsic, 2, 31
island, 49, 50

L

lasers, 2
lattice, 4, 18, 20, 29, 32, 33, 39, 40
lattices, 30
law, 29
light beam, 12
limitation, 33
linear, 3, 12, 13, 20, 36, 47
location, 53
luminescence, 21, 23, 26, 63
lying, 53

M

magnetic, vi

manufacturing, 2
matrix, 5, 11, 14, 35, 36, 37, 38, 64
MBE, 2
measurement, 45, 50, 60
microwave, 2
mixing, 3, 13
mobility, 54, 55
MOCVD, 2
modeling, 5
models, 9
modulation, 33
mole, 15, 70
molecular beam, 2, 71
molecular beam epitaxy, 2, 71
molecular-beam, 70
momentum, 14
monochromator, 28, 31
motion, 3

N

NATO, 68
New York, v, vi, 68, 69
noise, 2
normal, 3, 12, 13
n-type, 3, 12, 13, 45, 49, 56, 60

O

observations, ix
operator, 10, 37
optical, ix, 1, 3, 11, 12, 13, 49, 63, 71
optical absorption coefficient, 4
optical properties, 1, 4, 49
orbit, 9, 11, 17, 19, 20, 35, 71
oscillator, 12

P

parabolic, 11
parameter, 3, 29, 31
partition, 18, 20
passivation, 2
periodic, 5, 28, 33, 49, 69

periodicity, 27, 28, 32, 49, 50
perturbation, 8
perturbation theory, 8
phonon, 3
photoabsorption, 13
photodetectors, 3, 63, 71
photoemission, 3
photolithography, 49
Photoluminescence, 21, 23, 47, 70
photoluminescence spectra, 47
photon, 1, 13, 59
physics, 4
Planck constant, 10
plane waves, 5, 36
play, 30
polarization, 3
population, 25, 60
probability, 36, 37
program, 22, 32
p-type, ix, 3, 9, 13, 15, 41, 46, 49, 51, 54, 58, 60, 63, 64, 70, 71

Q

quantization, 7
quantum, ix, 1, 2, 3, 5, 6, 7, 9, 11, 12, 13, 14, 15, 21, 22, 26, 28, 35, 36, 37, 39, 40, 41, 42, 59, 63, 64, 70, 71
quantum well, 2, 3, 4, 5, 6, 7, 9, 11, 12, 13, 14, 15, 22, 28, 35, 36, 37, 39, 40, 41, 42, 59, 64, 70, 71

R

radiation, 1, 2, 3, 12, 13
random, 12, 32
range, 4, 28, 31, 55, 56
recombination, 23
reconstruction, 15
red shift, 22, 63
reflection, 1, 27, 29, 38, 44, 49
relationship, 24, 30, 47, 57, 63
relaxation, 28
resistance, 60

resolution, 27, 28, 33, 34, 63
room temperature, 1, 45
roughness, 23, 33

S

sample, 12, 15, 21, 23, 24, 30, 33, 34, 44, 45, 59, 60, 61
satellite, 2, 28, 29, 30, 31, 32, 33, 44, 47, 63
scattering, 14
Schrödinger equation, 5, 7, 8
searching, 7
semiconductor, 1, 2, 4, 5, 7, 8, 13, 17, 27, 35, 39, 71
semiconductors, 2, 9, 13, 15, 17, 20, 68, 69
sensing, 2
separation, 2, 29, 30, 39, 44
series, 3, 5
services, vi
silicon, 68
simulation, 28, 32, 33, 64
sites, 30
software, 55
spectrum, 1, 33, 45, 47
speed, 2
spin, 9, 11, 17, 19, 20, 35, 38, 39
stiffness, 39
strain, ix, 4, 9, 10, 19, 20, 22, 30, 34, 35, 38, 39, 40, 41, 42, 43, 44, 45, 46, 63, 64
strains, 42
strength, 11, 12
stress, 39
substitution, 30
superlattice, 30, 32, 33, 53
superlattices, 17, 27
surveillance, 3
symmetry, 13, 60, 61

T

Taiwan, 49

TEM, 33, 63
temperature, ix, 1, 24, 25, 26, 27, 45, 52, 53, 55, 56, 61, 63, 64, 70
temperature dependence, 26
terminals, 8
thermal equilibrium, 57, 58
three-dimensional, 11, 53
TIR, 12
TMC, 49
transfer, 5, 11, 37, 38, 64
transformation, 45
transition, 2, 3, 11, 14, 22, 25, 42, 43, 45, 46, 56, 59, 63, 64, 71
transitions, 1, 3, 11, 12, 13, 14, 25, 26, 27, 38, 45
transmission, 27, 33, 34, 36, 37, 38, 39, 40, 44, 53
transmission electron microscopy, 33
transport, 53, 56
tunneling, 4, 52, 53, 56, 64
two-dimensional, 2, 11, 53

U

uniform, 2, 33

V

valence, ix, 1, 3, 8, 9, 11, 13, 14, 17, 19, 20, 22, 23, 35, 36, 39, 40, 41, 57, 58, 64, 71

values, 20, 22, 24, 26, 31, 32, 39, 42, 44, 45, 47, 60
vapor, 2
variation, 41, 47
vector, 9, 10, 13, 36
velocity, 53, 54

W

wave vector, 9, 10, 13, 36, 37
waveguide, 45
wavelengths, 13, 41, 42, 43, 45
wells, ix, 3, 4, 5, 7, 11, 12, 13, 15, 21, 22, 23, 28, 29, 30, 33, 38, 41, 42, 44, 47, 57, 58, 59, 60, 61, 62, 63, 64, 71

X

x-ray diffraction, 27, 28, 63
XRD, 33, 44

Y

Y-axis, 32
yield, 2, 32

Z

zinc, 49